SCIENCE, TECHNOLOGY, AND DEMOCRACY

SUNY series in Science, Technology, and Society
Sal Restivo and Jennifer Croissant, editors

List of Titles

SCIENCE, TECHNOLOGY, AND DEMOCRACY

EDITED BY
Daniel Lee Kleinman

STATE UNIVERSITY OF NEW YORK PRESS

Science, technology, and
democracy /

Cover photos reprinted courtesy of State Historical Society of Wisconsin.
Top: Whi (X3) 52512 CF 58. Photograph by Clarence P. Schmidt.
Bottom: Whi (X3) 36467 CF 6725. Photograph by Del Ankers.

Published by
State University of New York Press, Albany

For information, address State University of New York Press,
90 State Street, Suite 700, Albany, NY 12207

Production by Cathleen Collins
Marketing by Anne M. Valentine

Library of Congress Cataloging in Publication Data

Science, technology, and democracy / edited by Daniel Lee Kleinman.
        p. cm. — (SUNY series in science, technology, and society)
    Includes bibliographical references and index.
    ISBN 0-7914-4707-3 (alk. paper) — ISBN 0-7914-4708-1 (pbk. : alk. paper)
    1. Science—Social aspects. 2. Technology—Social aspects. I. Kleinman, Daniel
Lee.   II. Series
Q175.5.S3728 2000
303.48′3—dc21

                                                            99-087563

10 9 8 7 6 5 4 3 2 1

FOR MY PARENTS
AND
ONCE AGAIN FOR
FLORA AND SUSAN

Contents

Acknowledgments

While my interest in the relationship between democracy and expertise is longstanding, the origins of this book are more recent. The early rumblings of the so-called science wars fueled my desire to write on the topic. Discussions with several people stimulated that writing and ultimately this project. Over a sixteen-plus hour drive between Atlanta and Madison, Wisconsin in the spring of 1995, Kristin Garrision and I gabbed about the politics of expertise, and it was a follow-up phone call from her some months later that led me to write a piece for *The Chronicle of Higher Education*. Discussions with Jill Dolan, strolling one autumn through Washington, D.C., helped clarify my thoughts and solidified my commitment. Over the years since the onset of the science wars, discussions with Allen Hunter, Scott Frickel, Jack Kloppenburg, Joan Sokolovsky, and Steven Vallas provided me with important sustenance. I am grateful to all of them.

This volume itself is an outgrowth of a session I organized at the 1997 annual meetings of the American Association for the Advancement of Science in Seattle (which was, in turn, the result of a promise I made in *The Chronicle*). Steve Schneider, one of the participants in the session, encouraged me to edit a collection. Several of the participants in that session are represented here. Other commitments prevented two of the panelists— Martha Crouch and Rustum Roy—from writing essays for this volume. Their talks in Seattle were thought-provoking, and I am sorry their ideas are not represented in this collection. Still, I am pleased that the authors represented here were able to contribute their work, and I am proud to have my name associated with such an impressive and astute group.

Of course, there would be no book without a publisher. I thank the series editors, Sal Restivo and Jennifer Croissant, and the acquisitions editors, James Peltz and Dale Cotton, at SUNY Press for their interest in the project. James, Dale, and Cathleen Collins provided crucial guidance as I navigated the publication process, and I am grateful to all of them. Thanks are due as well to the three reviewers for SUNY Press who read and provided support for the volume and useful comments on particular essays. At Georgia Tech, I have always been able to rely on superb logistical and administrative support from Denise Marshall, Rudy Paratchek, and Candy Snipes. All three have my thanks. Finally, I received the contract for this volume while a visiting fellow at Clare Hall (University of Cambridge) in 1998 and put the final touches on the volume while back again in 2000. Clare Hall is a wonderful place to work, and its members and staff have my appreciation.

I leave the most important for last. My parents, Barbara and Gerald Kleinman, provided the environment in which my early commitment to democracy was fostered. My daughter, Flora Berklein, has continued to show me how much more pleasurable doing scholarship is when it is balanced with creating Lego models, drawing pictures, building sand castles, and swimming. Finally, over the years, Susan Bernstein, has been a source of stimulating discussion and an appreciative consumer of my cooking. This book is for all of them.

Introduction

DANIEL LEE KLEINMAN

When I submitted the idea for this volume to SUNY Press, the title I proposed was *Beyond the Science Wars: Science, Technology, and Democracy.* The series editors suggested that the first part of the title be dropped. I assume that they were thinking of the shelf life of the book. In the years to come—whether it be two or ten years down the road—if the history of the so-called science wars and even the term itself are lost to collective memory, the contributions in this collection will still be valuable, and it would be a shame if the mere title led potential readers to overlook the book. Nevertheless, the timing of this collection's genesis is not coincidental. It should be understood in relationship to that often disturbingly hyperbolic dispute between a small group of scientists and so-called critics of science—a wide range of activists and scholars in the social sciences and humanities. It was the "science wars" that prompted me to put this book together, and I believe it is the ongoing reverberations of this discursive sparring and the related social conditions in which we find ourselves that make the contributions to it particularly vital now. Thus, I provide my assessment of the science wars as a frame for this volume.

Commentators agree that the origins of the "science wars" can be found in the 1994 publication of Paul Gross and Norman Levitt's *Higher Superstition: The Academic Left and Its Quarrels with Science.* Lauded by many scientist-reviewers and sharply criticized by a large number of scholars in the humanities and social sciences, Gross, a life scientist, and Levitt, a mathematician, spend nearly 300 pages in an admittedly polemical mode (1994, p. 14) examining what they view as the "proliferation of distortions and exaggerations about science" (1994, p. 7) found in a wide array of work in the social sciences and humanities. Gross and Levitt's stated fear is that this work threatens to "poison" the contemporary university (1994, p. 7), and they suggest that the attitudes found in it exist beyond the academy, in the environmental movement in particular (1994, p. 12).

Following on the successes of their book, Gross and Levitt and others organized a conference in the summer of 1995 entitled "The Flight from Science and Reason." Sponsored by the New York Academy of Sciences, the targets of conference participants were, if the edited collection derived from the gathering is any indication (Gross, Levitt, & Lewis, 1996), much broader than *Higher Superstition*. Sociologists of science and environmentalists were verbally tarred, but so were advocates of alternative medicine, health maintenance organizations, literary critics who challenge canonical thinking, and scholars who study gender and education. The supposed crime of which all of these people were guilty is nothing less than having lost their sense—taken a "flight from science and reason."

There was a flurry of activity in the months that followed—articles in periodicals, discussions on the Internet, and talk-show banter. But still, the "science wars" captured the imagination of only a limited public. Then on May 18, 1996—what must have been an extraordinarily slow news day—just below the fold in the *New York Times* a headline ran "Postmodern Gravity Deconstructed, Slyly" (Scott, 1996) and below the headline was a representation of the cultural studies journal, *Social Text*, until then relatively obscure beyond limited pockets of the academy. The *Times* told the story of a "hoax" perpetrated by a New York University physicist, Alan Sokal. He had written a gibberish blend of "postmodern philosophy" and physics theory, submitted it to *Social Text*, and the journal had accepted it for publication (Sokal 1996a). Sokal claimed that his "experiment" revealed the bankruptcy of much recent work by humanists and social scientists who study science. Importantly, he claimed these scholars do not understand the science on which they draw or which they analyze (Sokal, 1996b; Sokal & Bricmont, 1998).[1]

In the days and weeks that followed, analyses and responses to the so-called Sokal Affair found their way into such general circulation periodicals as the *New York Times* (Fish, 1996) and the *Village Voice* (Willis, 1996). Talk show hosts repeated Sokal's assessment, and noncognoscenti audiences were left to think that the university is filled with ignorant and dangerous scholars.[2]

The science wars have not faded with the Sokal affair. The year 1997 saw publication of a 42-essay volume, which originated in the "flight from science and reason" conference (Gross, Levitt, & Lewis, 1997). And Sokal has managed to remain in the spotlight with the 1997 publication in France of his co-authored book, *Intellectual Impostures* and its release and publicity tour in England in the summer of 1998.[3] In this book, Sokal and his coauthor Jean Bricmont attempt to "draw attention to a relatively little-known aspect [of postmodernism], namely the repeated abuse of concepts and terminology coming from mathematics and physics" (Sokal & Bricmont, 1998, p. 4).[4]

But if the academic field from which much of the work under attack comes—science studies—truly threatens to bring disrepute on the university and to spill over into the broader public arena, misleading citizens and poisoning public opinions about the scientific enterprise, why were the science wars just initiated in the mid-1990s? Significantly, the pathbreaking work that asked analysts of science to examine their views about how the practice of science really operates—studies that explored knowledge as a social construction—was published in the mid-1970s and early 1980s (cf. Barnes, 1974; Bloor, 1976; Latour & Woolgar, 1979; Knorr-Cetina, 1981). Furthermore, as Dorothy Nelkin has noted, science and scientists have been the butt of criticisms before, but their collective response lacked the current intensity. Scientists were unable to organize collectively in the 1970s to turn away efforts to promote "scientific creationism." In the 1980s, scientists generally left it to their directly affected colleagues to challenge animal rights activists' opposition to animal experimentation, and there was no high profile mobilization by scientists to support ongoing fetal research (Nelkin, 1996a, p. 94).

The intensity of reaction to science studies scholars and others who do not unquestioningly accept the authority of science is explained, I believe, by the decisive changes in the world in which we live. In the years from World War II through 1950 a pitched battle was fought in the United States over how research policy would be made in the period after the War. On one side was a group of populists led by a New Deal Senator from West Virginia, Harley Kilgore. Kilgore and his allies argued for a central agency that would be involved in coordinating the government's involvement in science and would include representatives from a wide array of social interests in the organization's decision making. Science elites led by Vannevar Bush argued for a more narrowly defined agency, an organization controlled by scientists and focusing primarily on support of so-called basic research. In exchange for autonomy and control over resources to support scientific research, the scientists promised science-generated progressive improvements in national social and economic well-being. In the event, Bush and his allies won the day (Kleinman, 1995a; Kleinman & Solovey, 1995). And until relatively recently that settlement—often called a social contract with science—has fundamentally defined the scientific community in the United States and the relationship between science and the citizenry. Throughout the Cold War, massive amounts of money flowed into university laboratories supporting research and training efforts that helped promote a picture of vibrant democracy in contrast to the totalitarian world of our Soviet adversaries. Decisions about the utilization of these resources were largely in the hands of certified scientists, and more generally, "technical matters" were considered the realm of certified experts alone.

But the Cold War is over. There is no Soviet Union, and while the fiscal crisis in the United States appears to be over, politicians across the spectrum seem committed to leaner federal budgets. And with the nation facing more forceful economic competition from countries around the world than might have been foreseen in the heady days after World War II, new federal priorities are under debate. Under these conditions, the postwar arrangements for science are no longer stable and the future direction of federal funding for science is uncertain (Kleinman, 1995b; Nelkin, 1996a, 1996b).[5] In the view of some, even recent modest budget increases for federal science agencies are not sufficient to support an increasingly capital intensive enterprise. In this context, large projects, like the Superconducting Supercollider, are found on the chopping block. Hoping to avoid having their priorities set by politicians, a few years back, some scientist groups engaged in their own priority setting exercises (Kleinman, 1995a, p. 191). And a distressed head of the American Association for the Advancement of Science found it necessary in 1991 to publish a study reporting the results of an informal survey he undertook. Leon Lederman found his colleagues' morale universally low and concluded that the source of discontent was the inadequate support of academic scientific research (Lederman, 1991).

As sufficient funding from the federal government for university science becomes increasingly questionable, scholars turn to industry for support, and as one analyst suggests, "In a climate of intense competition for patents and research funds, incidents of fraud, falsification of scientific evidence, and misconduct have proliferated" (Nelkin, 1996b, p. A52). Whether such incidents have, indeed, increased in number over the past decade or two is an open question, but it is clear that when such cases make it into the spotlight (cf. Kevles, 1998) they lead to greater regulation of the scientific community, a kind of intervention with which scientists are neither comfortable nor familiar.[6]

Changes have not only occurred at the policy-level, but in society at large as well. Not only did the scientific community make promises after World War II, but it kept them. Scientists delivered weapons systems that kept us safe during the Cold War, produced "Better Living Through Chemistry," and found cures for devastating diseases. In a rapidly growing economy in an optimistic age, scientists were considered crucial to the widespread realization of the American Dream.

But the environment slowly changed. Many people realized that although there were numerous successes, the scientific products of the postwar decades were not uniformly beneficent—a realization deepened by such high-profile disasters as those at Love Canal and Three Mile Island. A range of local and national political movements challenged the right of scientists to work without public scrutiny. Not only people on the political margins, but mainstream Americans criticized potential environmental and health hazards from toxic-waste dumps to nuclear power plants, for example, and threats posed to the family farm by agricultural technologies that raise costs or otherwise encourage the consolidation of holdings.[7]

It is this environment that I believe explains the timing and perhaps the form of the science wars. Indeed, some members of the scientific community appear to assume a transparent link between what they view as "flights from science and reason" and the changing science policy environment. In his 1998 candidate statement for the American Association for the Advancement of Science board of directors, Lewis Branscomb says:

> There is a movement to deny the rationality of public decisions based on science by claiming that all science is "socially constructed." Congress now demands that research agencies quantify the beneficial outcomes from their research investments. What can be done? (Branscomb, 1998, p. 4)

In the first sentence here, Branscomb dismisses by caricature recent work in science studies. His second sentence bemoans the breakdown of the postwar social contract for science, which means increased government oversight. He does not explicitly link the two, apparently assuming the relationship is obvious.

Although at some level and in some cases, the concerns of Gross, Levitt, Sokal, and others may be legitimate, there can be little doubt that these disputes constitute an effort to reinforce a crumbling boundary: a wall that divided scientists and lay citizens, a barrier that legitimated scientists' autonomy on expert matters and dictated citizen silence.[8] In this context, surveys that report overwhelming public support for science and scientists may be of little comfort to those who seek to maintain a hard and fast boundary between expert and layperson. For these same surveys suggest that by traditional measures substantial portions of the U.S. populous are "scientifically illiterate" (Lawler, 1996; see also Freudenburg, 1996). Traditional science literacy may prompt a fascination with milestones in research, but also a recognition that only scientists are equipped to deal with the complexities of the "natural world" (Kleinman & Kloppenburg, 1991). By contrast, if this "illiterate" public turns to scholars and others whose work points to the fundamentally social character of science, work that may blur the boundary between expert and citizen, for understanding, they may be led to believe that they are capable of playing a role, or entitled to intervene, in the realm of science. Scientists used to the traditional rigid division between scientists and other citizens are likely to find this prospect worrying.

Indeed, as to the question of why waste energy attacking a collection of nerds who nobody listens to, Norman Levitt is quite clear: "The catchphrases that rebound in the lecture hall one year tend, for good or ill, to be heard at rallies the next" (1996, p. 30). If scholars have the nerve to challenge the authority of scientists, it will not be long, this reasoning suggests, before average citizens feel emboldened to do precisely the same. The flip side of this logic is equally true. In their focus on academic researchers, Gross, Levitt, and Sokal have asserted that humanists and social scientists should not criticize that which they do not understand. As Sokal and Bricmont put it: "The sensible conclusion . . . is that sociologists of science ought not to study scientific controversies on which they lack the competence to make an independent assessment of the facts . . ." (1998, p. 90).[9] It is only a short step from that claim to the contention that nonexperts have no right to intervene in the realm of science. Indeed, Gross and Levitt make that argument quite explicitly in an article they published in *The Chronicle of Higher Education*. According to Gross and Levitt, "Scientific decisions cannot be submitted to a plebiscite; the idea is absurd. Applied to science education, for example, letting people vote on what should be taught would give us countless schools in which 'creation science' would replace evolutionary biology" (1994b, p. B2). The image rendered here is of a mass uninformed public voting on matters in the absence of information or a commitment to understand the issues at stake.

But in the vast literature on science, technology, and democracy, I have found no one who advocates such an absurd situation. Importantly, a recent exercise has shown that "people who don't ordinarily keep abreast of scientific issues can quickly learn about their critical aspects" (Doble & Richardson 1992, p. 52), and several cases discussed in this volume confirm this finding. At the same time, during a period in which the impacts of science and technology are felt in the daily lives of citizens throughout the world, principles of democracy dictate that we at least consider the plausibility of increasing citizen involvement in the realm of science. It is this claim—and my belief that it is time to move beyond the hyperbole that have dominated the science wars—that led me to bring together in this book the work of an illustrious group of activists, scientists, and science studies scholars. I required that contributors accept no particular definition of democracy or citizen involvement in science. Instead, the premise is that in some fashion citizens can be involved in decisions concerning science and technology.

Overview

This collection is divided fairly evenly into two relatively distinct sections. In the first, authors describe and assess real life examples of specific cases of democratic participation in matters of science and technology. The essays in the second part of the book enter into the discussion from a different angle, pondering large questions. The contributors consider, for example, what constitutes expertise and how we can distinguish between efforts to democratize science and technology and contemplate what kinds of policies and institutions might help us move in the direction of increasing or improving citizen involvement in matters of science and technology.

In chapter 1, Steven Epstein uses the case of AIDS treatment activism in the United States as a tool to explore the conditions under which it is possible for grassroots activists to challenge the traditional social organization of expertise. Although activists were able to infiltrate the biomedical community and ultimately captured "seats at the table," playing important roles in, for example, developing clinical trial protocols, their story does not provide a simple road map for other activists interested in gaining access to research decision-making. Epstein clearly illustrates that it was the particular demographic profile of AIDS treatments activists (they were primarily white, male, and highly educated) as well as the movement infrastructure developed by earlier gay activists that provided the foundation for treatment activists' entrée into the world of biomedical science. Epstein speculates that lower-status social groups will have less luck in their efforts to gain voice in matters of biomedicine. In addition, Epstein points to the kind of unintended consequences that can befall movements seeking to influence biomedical research- and practice-related decision making. On the one hand, for example, such movements are likely to reproduce in their own organizations the expert-lay divide that they seek to undermine in the larger scientific arena; and, on the other hand, Epstein suggests that knowledgeable clinical trial participants may not abide by trial rules, thus muddying study results.

In chapter 2, Richard E. Sclove explores a very different kind of citizen participation. While AIDS treatment activism has extra-institutional origins and seeks to promote intimate lay involvement in research practice, the consensus conferences Sclove describes are vehicles for citizen influence in larger matters of science and technology policy. Sclove traces the history of consensus conferences in Denmark and their spread to other European nations and most recently to the United States. He details the practical mechanics of this organizational form and provides evidence suggesting that laypeople in these environments can grapple intelligently with highly technical matters and that the reports produced by these bodies can influence business decision-making. In the last portion of his contribution, Sclove reflects on the first U.S. consensus conference—an effort in which he was intimately involved. He suggests that the success of this first-of-its-kind effort should make us optimistic about the potential viability and efficacy of regularizing use of this decision-making model.

The existing writings of Epstein and Sclove already constitute important sources for readers interested in issues of science, technology, and democracy. In chapters 3 and 4, we hear from new voices. Neva Hassanein's contribution—chapter 3—adds a site not often discussed in debates about knowledge and democracy. Drawing on data collected over two years in Wisconsin, Hassanein discusses and analyses two farmer-to-farmer networks, which in their distinct ways illustrate the importance of "local knowledge" and of sharing that knowledge. Hassanein begins by illustrating the ways in which the social organization of power has shaped agricultural science and limited farmer input into agricultural science agendas and research. She goes on to explore the two networks she studied: one that works to develop and circulate knowledge of relevance to a sustainable agricultural practice called rotational grazing, and the other a network of women farmers interested in sustainable agriculture who work to democratize knowledge about social relations in agriculture. Hassanein suggests that the successes of the networks she describes constitute an important challenge to "the inequitable power relations characteristic of the dominant system of agricultural knowledge production and distribution."

In the final chapter in the book's first section, Louise Kaplan offers a history of citizen involvement in decision making on issues related to nuclear power, nuclear weapons, and nuclear waste disposal. Kaplan traces the "slow motion" transformation of the citizenry near the Hanford nuclear facility in Washington state from passive bystanders who accepted the professional qualifications and consequent assurances of experts concerning the safety of the Hanford site to informed and outraged participants who illustrated their competence in discussions about regulation of the Hanford facility. Kaplan's study points to the power of the idea of scientists as neutral experts and the reality that the perspectives of experts reflect a distinctive set of biases. In addition, like other chapters in the section, Kaplan's essay challenges naysayers who assert that laypeople cannot intelligently grapple with highly technical matters.

Daniel Sarewitz's contribution leads off the second section of this volume. Sarewitz explores why an Enlightenment view of science lies in tension with the ultimately uncontrollable character of nature and the similar lack of predictability and controllability that gives democracy its vitality. Sarewitz begins by describing the pervasive impact of the Cold War on science practice and policy in the United States and then explores what he calls the "Enlightenment Program." The core of the essay is an outline of eight diverse science-related problems that we currently face. Sarewitz suggests these are intimately linked to science practices shaped by the particular manifestation of the Enlightenment view that guided science in the United States during the Cold War. He concludes by arguing that "the organizational structure and knowledge products of today's [scientific] enterprise are often not suited to addressing . . . [these problems] productively" and calls for a new approach to federal science policy and practice better suited than the current system to promoting human needs and well-being.

In chapter 6, Stephen Schneider does not question the view, supported by the contributions in part I, that laypeople are capable of understanding complex scientific arguments and the character of debates between scientists. He does suggest, however, that such a level of understanding demands a kind of commitment that few lay citizens are likely to make. Consequently, Schneider proposes the creation of a "meta-institution" that will provide citizens and their elected representatives with assistance in evaluating scientific credibility. Citizens would witness the proceedings of this organization, but membership on the body would be dominated by persons nominated by scientific societies. Without such a body, Schneider fears citizens put off by "baffling technical brouhaha" are likely to abdicate to experts their role in the "value-laden" policy selection process. It is worth noting that Schneider assumes something questioned by several other contributors to this volume: that it is possible to separate the value and factual components of technical matters. Schneider sees a role for lay citizens in the value aspects of technical matters. By contrast, in differing ways, the other contributors to the collection point to the inseparability of fact and value and suggest a role for lay citizens in what Schneider might take to be the technical/factual core of science.

Like Daniel Sarewitz, in her chapter, Sandra Harding asks us not to limit ourselves to an Enlightenment model of science. Harding's critique focuses not on the issue of control, but instead on the problematic claim that social and political neutrality can and should characterize sciences' internal, cognitive, technical core. She suggests that we need to be attentive to "how social and political fears and desires get encoded in that purportedly purely technical, cognitive core of scientific projects." In particular, Harding examines the universality ideal embedded in much contemporary science and suggests that this underlying principle leads to the devaluation of cognitive diversity, legitimates the acceptance of less well-supported claims over stronger ones in some instances, and creates blindness to some of the most cogent criticisms of particular scientific contentions. Still, Harding does not want to abandon the universality ideal entirely, but instead wishes to retain its valuable components and reconceptualize others in a way that will permit us to realize important democratic ideals.

In the final chapter of book, I contribute to the debate on citizen involvement in the realm of technoscience. I suggest that discussions of democratic involvement in science and technology are often marred by lack of clarity and consequent misunderstanding. In an effort to bring lucidity to these exchanges, I outline several dimensions across which it is possible to distinguish instances of democratized science. Drawing on cases considered in this volume and elsewhere, I contend that laypeople can grasp the subtle content, difficult concepts, and methodological complexity of science, and consequently, I suggest that this is not a valid basis for a priori rejection of efforts to democratize science. I suggest that the real obstacles to the democratization of science are rooted in widespread social and economic inequalities and an unexamined commitment to expert authority. I conclude by providing some rudimentary proposals for overcoming these hurdles.

At the Starting Line

After Alan Sokal's hoax was made public, discussion of what is at stake in the science wars—some of it quite useful—proceeded faster and more furiously than it had until that point. In venues from the Internet to letters to the editor, discussants—from cultural studies celebrities to unknown graduate students— expressed their views. In his letter to the editors of *Lingua Franca*, Rutgers University's George Levine made a comment that might usefully serve as a guiding premise for this collection. "The key," suggested Levine, ". . . is that the public should have a responsible *and* intelligent relationship to science" (1996, p. 64). I hope this volume contributes to such a relationship. As such, it should constitute not the last word on the science wars or on the relationship between citizenship and science and technology, but rather, I hope the chapters that follow will prompt productive dialogue.

Notes

1. See the series of responses to Sokal in *Lingua Franca*, July/August, 1996.

2. For a perceptive analysis of Sokal's "experiment" and why it does not show what he claims, see Hilgartner (1997).

3. I was surprised to see *Intellectual Impostures* among the "best sellers" when I happened into a Waterstones bookstore in Cambridge, England in January 2000. The science wars continue.

4. It is my view that the written work by the science warriors does not provide a fair or damning critique of science studies. See Kleinman (1995b) and Kleinman (1999). See also Lewenstein (1996) and Guston (1995).

5. Readers wishing to follow the recent history of federal budget battles as they relate to science and technology should turn to back issues of *Science* magazine.

6. Congressional involvement in discussions about (mis)conduct in science can be traced in *Science* magazine.

7. This paragraph and part of the previous paragraph are taken from Kleinman (1995b).

8. In the epilogue to his recent book, Thomas Gieryn (1999) points out that it is not only scientists in the science wars who are engaged in boundary construction, defense, and expansion. Many in science studies are involved in the same exercise.

9. The question is who decides what counts as competence. Levitt, Gross, Sokal, and Bricmont would presumably assert that only certified scientists are qualified to assess the competence of nonscientists. If this is the case, any criticism of science can only occur on scientists' terms.

Bibliography

Barnes, B. (1974). *Scientific knowledge and sociological theory.* London: Routledge and Kegan Paul.

Bloor, D. (1976). *Knowledge and social imagery.* London: Routledge and Kegan Paul.

Branscomb, L. M. (1998). Statement for candidate of Board of Directors. Election 98. Washington, DC: American Association for the Advancement of Science.

Doble, J., & Richardson, A. (1992, January). You don't have to be a rocket scientist. . . . *Technology Review,* 51–54.

Fish, S. (1996, May 21). Professor Sokal's bad joke. *The New York Times,* p. A11.

Freudenburg, W. R. (1996). Gross, Levitt, waste wars and witches: Diversionary reframing and the social construction of superstition. *Technoscience, 9*(2), 26–29.

Gieryn, T. F. (1999). *Cultural boundaries of science: Credibility on the line.* Chicago: University of Chicago Press.

Gross, Paul R. and Levitt, N. (1994). *Higher superstition: The academic left and its quarrels with science.* Baltimore, MD: John Hopkins University Press.

Gross, P. R., Levitt, N., & Lewis, M. W. (Eds.). (1996). *The flight from science and reason.* New York: New York Academy of Sciences.

Guston, D. H. (1995). The flight from reasonableness. *Technoscience, 8*(3), 11–13.

Hilgartner, S. (1997). The Sokal affair in context. *Science, Technology, and Human Values, 22*(4), 506–522.

Kevles, D. J. (1998). *The Baltimore case: A trial of politics, science, and character.* New York: W. W. Norton and Company.

Kleinman, D. L. (1999). Defining disagreements: From intolerance to civil dialogue in the science wars. *Configurations, 7,* 101–108.

Kleinman, D. L. (1995a). *Politics on the endless frontier: Postwar research policy in the United States.* Durham, NC: Duke University Press.

Kleinman, D. L. (1995b, September 29). Why science and scientists are under fire—and how the profession needs to respond. *The Chronicle of Higher Education,* B1–B2.

Kleinman, D. L., & Kloppenburg, J. (1991). Aiming for the discursive high ground: Monsanto and the biotechnology controversy. *Sociological Forum, 6*, 427–447.

Kleinman, D. L., & Solovey, M. (1995). Hot science/Cold war: The National Science Foundation after World War II. *Radical History Review, 63,* 110–139.

Knorr-Cetina, K. (1981). *The manufacture of knowledge: An essay on the constructivist and contextual nature of science.* Oxford: Pergamon.

Latour, B., & Woolgar, S. (1979). *Laboratory life: The social construction of scientific facts.* Beverly Hills, CA: Sage.

Lawler, A. (1996, May 31). Support for Science Stays Strong. *Science, 272,* 1256.

Lederman, L. (1991). *Science: The end of the frontier.* Washington, DC: American Association for the Advancement of Science.

Levine, G. (1996, July/August). Contribution to The Sokal hoax: A forum. *Lingua Franca,* 64.

Levitt, N. (1996). Response to Freudenburg. *Technoscience, 9*(2), 29–30.

Levitt, N., & Gross, P. (1994a). *Higher superstition: The academic left and its quarrels with science.* Baltimore, MD: Johns Hopkins University Press.

Levitt, N., & Gross, P. (1994b, October 5). The perils of democratizing science. *The Chronicle of Higher Education,* B1, B2.

Lewenstein, B. V. (1996, July 21). Science and society: The continuing value of reasoned debate. *The Chronicle of Higher Education,* B1.

Nelkin, D. (1996a). The science wars: Responses to a marriage failed. *Social Text, 14,* 1/2: 93–100.

Nelkin, D. (1996b, July 26). What are the science wars really about. *The Chronicle of Higher Education,* A52.

Scott, J. (1996, May 18). Postmodern gravity deconstructed, slyly. *The New York Times,* pp. 1, 11.

Sokal, A. (1996a). Transgressing the boundaries: Toward a transformative hermeneutics of quantum gravity." *Social Text, 14,* 1/2: 217–252.

Sokal, A. (1996b, May/June). A physicist experiments with cultural studies." *Lingua Franca,* 62–64.

Sokal, A., & Bricmont, J. (1998). *Intellectual impostures.* London: Profile Books, Ltd.

Willis, E. (1996, June 25). My Sokaled life. *Village Voice,* pp. 20, 21.

Part I

Citizen Participation in Action

1

Democracy, Expertise, and AIDS Treatment Activism

STEVEN EPSTEIN

In 1987, thousands of AIDS activists around the United States began confronting doctors, biomedical researchers, and federal health officials in visually arresting, angry, and provocative demonstrations. Although the targets varied, a lot of this new wave of AIDS activism focused on the organization and pace of research on AIDS treatments. The messages were not subtle. At one scientific forum, activists handed out cups of Kool-Aid as a prominent researcher came to the podium, likening the effects of his research methods on AIDS patients to that of cult leader Jim Jones on his followers in Jonestown, Guyana (Crowley, 1991, p. 40). When the Commissioner of the Food and Drug Administration (FDA) came to speak at a public forum in Boston in 1987, activists in the audience held wristwatches aloft ("FDA allows," 1988): for people with AIDS, these activists implied, time was running out. In October of that year, more than 1,000 demonstrators converged on FDA headquarters in Rockville, Maryland to "seize control" of what some labeled the "Federal Death Administration" (Bull, 1988).

Fast-forward to 1992: A subset of these same activists now sat as regular voting members on the committees of the AIDS Clinical Trials Group (ACTG), the entity established by the National Institutes of Health (NIH) to oversee all federally funded, AIDS clinical research. Serving alongside the most prominent AIDS researchers in the country (including the one who had been compared to Jim Jones), activists now worked with scientists to determine the most profitable research directions, debate research methodologies, and allocate research funds. Activists also served on institutional review boards at research hospitals around the country, evaluating the methods and

15

ethics of clinical trials of AIDS drugs. At conferences, where once they had shouted from the back of the room, activists now chaired sessions. And their publications, like San Francisco–based *AIDS Treatment News,* had become routine sources of information about AIDS therapies for many doctors around the world. The aptitude of AIDS treatment activists in understanding such matters as the stages of viral replication, the immunopathogenesis of HIV, and the methodology of the randomized clinical trial was widely acknowledged by prominent experts. As Dr. John Phair (1994), a former chair of the Executive Committee of the AIDS Clinical Trials Group, commented in 1994: "I would put them up against—in this limited area— many, many physicians, including physicians working in AIDS [care]. They can be very sophisticated."

The unusual social movement trajectory that I have sketched—with the contrast between banging on the doors of biomedicine in 1987 and sitting at the table by 1991—is interesting for all sorts of reasons. Here I would like to focus on the *politics of knowledge and expertise:* How did a grassroots movement produce a cadre of activist-experts? More generally, in what circumstances can laypeople challenge hierarchies of expertise and participate effectively in processes of scientific knowledge-production—or transform such processes? How do they gain entry to these privileged domains, and what are the consequences, intended or unintended, of these kinds of incursions?[1]

What I want to argue here is that activist movements, through amassing different forms of credibility, can in certain circumstances bring about changes in the epistemological practices of science—our ways of knowing the natural world. Nothing guarantees that such changes will be useful in advancing knowledge or in curing disease, but in this case, I want to suggest, lay participation in science had some tangible benefits, though not without risks. This is a surprising finding, and one that is, of course, at variance with popular notions of science as a relatively autonomous arena with high barriers to entry. And it runs counter to the view that many might normally voice—that science must be safeguarded from external pressures in order to prevent the deformation of knowledge.[2]

By scientific "credibility," I refer to the capacity of claims-makers to enroll supporters behind their claims and present themselves as the sort of people who can give voice to scientific truths.[3] I understand credibility as a form of authority that combines aspects of power, legitimation, trust, and persuasion (Weber, 1978, pp. 212–254). However, this case differs from other sociological studies of scientific credibility by suggesting the diversity of the cast of characters who strive for credibility in scientific controversies, and the variety of routes by which credibility is made manifest. More typically in science the attestations of credibility are recognizable markers like academic degrees, research track records, institutional affiliations, and so on. In the case of AIDS research, when we examine the interventions of laypeople and activist movements, we find a multiplication of the successful pathways to the establishment of credibility, a diversification of the personnel beyond the formally credentialed, and hence more convoluted routes to the resolution of controversy and the construction of belief.

This study also presumes a particular historical moment in which, perhaps especially in the United States, popular attitudes toward science and medicine are highly polarized: a deep faith on the part of the public proceeds hand in hand with skepticism and disillusionment. The emergence of a new epidemic disease, in a society inclined to consider itself as having advanced beyond such mundane risks, had the effect of amplifying this ambivalence. When experts appeared unable to solve the problem of AIDS, the resulting disappointment created space for unconventional voices. Therefore, the study of credibility in AIDS research must also be a study of the public negotiation of the "credibility crisis" surrounding biomedical science—and a study of how the boundaries around scientific institutions become more porous, more open to the intervention of outsiders, precisely in such moments. Particularly when scientific credibility is in crisis, science may become the site of wide-ranging credibility struggles. Again perhaps especially in the United States, interventions by outsiders may get organized in the form of full-fledged social movements.

Origins of the AIDS Treatment Activist Movement

The U.S. AIDS treatment activist movement is best conceived as a subset of a much larger, but considerably more diffuse, "AIDS movement" that dates to the early years of the epidemic; that encompasses a wide range of grassroots activists, lobbying groups, service providers, and community-based organizations; and that now represents the diverse interests of people of various races, ethnicities, genders, sexual preferences, and HIV "risk behaviors." The AIDS movement has engaged in manifold projects directed at a variety of social institutions, including the state, the church, the mass media, and the health care sector (Altman, 1994; Cohen, 1993; Corea, 1992; Crimp & Rolston, 1990; Elbaz, 1992; Emke, 1993; Geltmaker, 1992; Indyk & Rier, 1993; Patton, 1990; Quimby & Friedman, 1989; Treichler, 1991; Wachter, 1991)—though at times it has been less concerned with achieving institutional change than with posing general challenges to cultural norms (Gamson, 1989).

In its emergence and mobilization, the AIDS movement was a beneficiary of "social movement spillover" (Meyer & Whittier, 1994): it was built on the foundation of other movements and borrowed from their particular strengths and inclinations. Most consequential was the link to the lesbian and gay movement of the 1970s and early 1980s (Adam, 1987; Altman, 1982,

1986). In the wake of fierce debates in the 1970s over whether homosexuality should be classified as an illness (Bayer, 1981), gay men and lesbians were often inclined toward critical or skeptical views of medical authorities (Bayer, 1985). More generally, it mattered that gay communities had preexisting organizations that could mobilize to meet a new threat; these community organizations and institutions also provided the face-to-face "micro-mobilization contexts" that are particularly useful in drawing individuals into activism (Lo, 1992). It mattered, too, that these communities contained (and in fact were substantially dominated by) white, middle-class men with a degree of political clout and fundraising capacity unusual for an oppressed group. And it was crucially important that gay communities possessed relatively high degrees of "cultural capital"—cultivated dispositions for appropriating knowledge and culture (Bourdieu, 1990). Within these communities are many people who are themselves doctors, scientists, educators, nurses, professionals, or other varieties of intellectuals. On the one hand, this has provided the AIDS movement with an unusual capacity to contest the mainstream experts on their own ground. On the other hand, it affords important sources of intermediation and communication between "experts" and "the public." Treatment activists themselves have tended to be science novices, but ones who were unusually articulate, self-confident, and well-educated—"displaced intellectuals from other fields," as Jim Eigo, a New York City treatment activist with a background in the arts, expressed it (Antiviral Drugs Advisory Committee, 1991, p. 50).

The fact that many lesbians (and heterosexual women) who would become active in the AIDS movement were schooled in the tenets of the feminist health movement of the 1970s (Corea, 1992; Winnow, 1992)—with its skepticism toward medical claims-making and insistence on the patient's decision-making autonomy (Boston Women's Health Book Collective, 1973; Fee, 1982; Ruzek, 1978)—also had important implications for the identity and strategies of the movement. Other activists, both men and women, had prior direct experience in social movements such as the peace movement (Elbaz, 1992, p. 72).

Central to the early goals of the AIDS movement was the repudiation of helplessness or "victim" status and the insistence on self-representation. "We condemn attempts to label us as 'victims,' which implies defeat, and we are only occasionally 'patients,' which implies passivity, helplessness, and dependence upon others. We are 'people with AIDS,'" read a widely reprinted manifesto of the New York–based PWA [People with AIDS] Coalition (PWA Coalition 1988). This is "self-help with a vengeance," as Indyk and Rier (1993, p. 6) nicely characterize it—an outright rejection of medical paternalism and an insistence that neither the medical establishment, nor the government, nor any other suspect authority would speak on behalf of people with AIDS or HIV.

AIDS activism entered a new and more radical phase in the second half of the 1980s, in the face of increasing concern about the inadequacy of the federal response to the epidemic, the stigmatization of people with AIDS or HIV, and the lack of availability of effective therapies for AIDS or its associated opportunistic infections and cancers. The year 1987 marked the birth, in New York City and then elsewhere around the country, of a new organization, called the AIDS Coalition to Unleash Power, but better known by its deliberately provocative acronym, ACT UP (Anonymous, 1991). A magnet for radical young gay men and women in the late 1980s, ACT UP practiced an in-your-face politics of "no business as usual." Adopting styles of political and cultural practice deriving from sources as diverse as anarchism, the peace movement, the punk subculture, and gay liberation "zaps" of the 1970s, ACT UP became famous for its imaginative street theater, its skill at attracting the news cameras, and its well-communicated sense of urgency. ACT UP groups typically had no formal leaders, and meetings in many cities operated by the consensus process.

ACT UP was only the most visible of a diverse set of groups that became interested in issues of medical treatment and research for AIDS around the United States in the mid- to late 1980s—the constellation of organizations that can be called the AIDS treatment activist movement. So-called buyers clubs, existing on the fringes of the law, supplied patients with unapproved or experimental treatments smuggled in from other countries or manufactured in basement laboratories. Project Inform, a San Francisco–based organization with a more conventional structure than ACT UP, emerged as an advocate for the use of such experimental therapies and evolved into a multifocal treatment advocacy organization with its own lobbying campaigns, publications, and educational projects. A range of grassroots treatment publications appeared, providing their readers with a rich mix of scientific information, political commentary, and anecdotes about treatments gleaned from patient reports (Arno & Feiden, 1992; Kwitny, 1992). *AIDS Treatment News*, the most well-known of these alternative publications, had been advocating for some time for greater attention to be paid to issues of drug research and regulation. "So far, community-based AIDS organizations have been uninvolved in treatment issues, and have seldom followed what is going on," wrote its editor, John James, a former computer programmer, in a call to arms in May 1986:

With independent information and analysis, we can bring specific pressure to bear to get experimental treatments handled properly. So far, there has been little pressure because we have relied on experts to interpret for us what is going on. They tell us what will not rock the boat. The companies who want their profits, the bureaucrats who want their turf, and the doctors who want to avoid making waves have all been at the table. The persons with AIDS who want their lives must be there, too.

To "rely solely on official institutions for our information," James bluntly advised, "is a form of group suicide" (James, 1986).

Gaining Credibility

Although treatment activists began, as I described, by implementing some highly confrontational modes of direct action, they always assumed that effective solutions to AIDS would have to come, in large measure, from doctors and scientists. Therefore, they resisted the notion—found, for example, in the animal rights movement (Jasper & Nelkin, 1992)—that the scientific establishment was "the enemy" in an absolute sense. "I wouldn't exaggerate how polite we were," Mark Harrington (1994), one of the leaders of ACT UP/New York's Treatment & Data Committee reflected:

> At the same time, I would just say that it was clear from the very beginning, as Maggie Thatcher said when she met Gorbachev, "We can do business." We wanted to make some moral points, but we didn't want to wallow in being victims, or powerless, or oppressed, or always right. We wanted to engage and find out if there was common ground.

How did this rapprochement proceed? In effect, activists (or some subset of them) accomplished an identity shift: They reconstituted themselves as a new species of expert—as laypeople who could speak credibly about science in dialogues with the scientific research community. I cannot here consider in detail the specific tactics that activists employed to construct their scientific credibility (see Epstein, 1995), but I would argue that four tactics were most important. First, activists acquired cultural competence by learning the language and culture of medical science. Through a wide variety of methods—including attending scientific conferences, scrutinizing research protocols, and learning from sympathetic professionals both inside and outside the movement—the core treatment activists gained a working knowledge of the medical vocabulary. Second, activists presented themselves as the legitimate, organized voice of people with AIDS or HIV infection (or, more specifically, the current or potential clinical trial subject population). Once activists monopolized the capacity to say "what patients wanted," researchers could be forced to deal with them in order to ensure that research subjects would both enroll in their trials in sufficient numbers and comply with the study protocols.[4] Third, activists yoked together methodological (or epistemological) arguments and moral (or political) arguments, so as to multiply their "currencies" of credibility. For example, activists insisted that the inclusion of women and people of color in clinical trials was not only morally necessary (to ensure equal access to potentially promising therapies) but was also scientifically advisable (to produce more fully generalizable data about drug safety and efficacy in different populations). Finally, activists took advantage of preexisting lines of cleavage within the scientific establishment to form strategic alliances. For example, activists struck alliances with biostatisticians in their debates with infectious disease researchers about appropriate clinical trial methodology.

A key victory for activists, at a time when many AIDS researchers remained deeply suspicious of the activist agenda, was the support of Dr. Anthony Fauci, prominent immunologist and AIDS researcher, and director of NIH's Office of AIDS Research. "Something happened along the way," Fauci told a reporter in 1989: "People started talking to each other. . . . I started to listen and read what [activists] were saying. It became clear to me that they made sense" (Garrison, 1989, p. A-1). Of course, Fauci and others may have deemed it strategic to incorporate activists into the process: As Fauci (1994) later commented, the assumption was that "on a practical level, it would be helpful in some of our programs, because we needed to get a feel for what would play in Peoria, as it were." Prominent academic researchers also acknowledged the gradual acquisition of scientific competence on the part of key activists. Dr. Douglas Richman (1994), an important AIDS researcher from the University of California, San Diego, described how Harrington of ACT UP/New York, in an early meeting with researchers, "got up and gave a lecture on CMV [cytomegalovirus] . . . that I would have punished a medical student for—in terms of its accuracy and everything else—and he's now become a very sophisticated, important contributor to the whole process."

As this encounter between different social worlds unfolded, activists pressed for more substantial degrees of inclusion in the NIH decision-making apparatus. Crucial decisions about clinical trials—which ones to fund, but also quite specific details about how the trials should be conducted, how the data should be analyzed, and which patients should be eligible to participate—were being made by the academic researchers who comprised the advisory committees of the ACTG. To the consternation of the researchers, activists demanded representation on these committees; when Fauci stalled, activists decided to "Storm the NIH." This demonstration, at the NIH campus in Bethesda, Maryland on May 21, 1990, proved to be another graphic media spectacle, like the FDA protest two years earlier (Hilts, 1990). Soon afterward, activists were informed that most ACTG meetings would be opened to the public, and that there would be a representative of the patient community, with full voting rights, assigned to each ACTG committee. In addition, by the early 1990s, activists acted as informal representatives to FDA advisory committees charged with evaluating new drugs, as appointed members of "community advisory boards" established by pharmaceutical companies, and as regular members of "institutional review boards" supervising clinical studies at hospitals and academic centers around the country. Activists began to have an important say in how studies were conducted, which patients were allowed into studies, how results from studies were evaluated, and which lines of research should be funded (Epstein, 1996; Arno & Feiden, 1992; Kwitny, 1992; Jonsen & Stryker, 1993).

Consequences

A defining moment was the publication in the *New England Journal of Medicine,* in November 1990, of an article by AIDS researcher Thomas Merigan (1990) of Stanford, entitled, "You *Can* Teach an Old Dog New Tricks: How AIDS Trials Are Pioneering New Strategies." Praising the new "partnership of patients, their advocates, and clinical investigators," Merigan proceeded to endorse precisely those methodological stances that activists had promoted. He argued, for example, that "all limbs [of a trial] should offer an equal potential advantage to patients, as good as the best available clinical care"; that no one in a trial should be denied treatment for their opportunistic infections; that trials should not be "relentlessly pursued as originally designed" when "data appeared outside the trial suggesting that patients would do better with a different type of management"; and that "the entry criteria for trials should be as broad as scientifically possible to make their results useful in clinical practice" (Merigan, 1990, p. 1341).

By pressing researchers to develop clinically relevant trials with designs that research subjects would find acceptable, activists helped to ensure more rapid accrual of the required numbers of subjects and to reduce the likelihood of noncompliance. And by working toward methodological solutions that satisfy, simultaneously, the procedural concerns of researchers and the ethical demands of the patient community, AIDS activists have, at least in specific instances, improved a tool for the production of scientific facts in ways that even researchers acknowledge. In this sense, AIDS activists' efforts belie the commonplace notion that only the insulation of science from "external" pressures guarantees the production of secure and trustworthy knowledge.

Those who were critical of lay participation in science were quick to suggest, and with some reason, that AIDS activists had muddied the waters of knowledge in their haste to see drugs approved. Yet any such assessment has to consider the larger picture. Absent the activists, what sort of knowledge strategies would have been pursued? Pristine studies addressing less-than-crucial questions? Methodologically unimpeachable trials that failed to recruit or maintain patients? Inevitably, there are risks inherent in the interruption of the status quo. But these must be weighed against all the other attendant risks, including those that may follow from letting normal science take its course while an epidemic rages.

One reason why this case is so important is because it is quite conceivable that these changes in the arena of AIDS research will have an enduring impact on biomedicine in the United States.[5] The past few years have seen a marked upsurge of health-related activism of a distinctive type: the formation of groups that construct identities around particular disease categories and assert political and scientific claims on the basis of these new identities. Just as the AIDS movement drew on the experiences of other movements that preceded it, now its own tactics and understandings have begun to serve as a model for a new series of challengers.

Most notably patients with breast cancer, but also those suffering from chronic fatigue, environmental illness, prostate cancer, mental illness, Lyme disease, Lou Gehrig's disease, and a host of other conditions, have displayed a new militancy and demanded a voice in how their conditions are conceptualized, treated, and researched (Barinaga, 1992; Kingston, 1991; Kroll-Smith & Floyd, 1997). These groups have criticized not only the quality of their care, but also the ethics of clinical research ("Are placebo controls acceptable?") and the control over research directions ("Who decides which presentations belong on a conference program?"). While not every such group owes directly to AIDS activism, the tactics and political vocabulary of organizations like ACT UP would seem, at a minimum, to be "in the wind." (Could one imagine, before the AIDS activist repudiation of "victimhood," people with muscular dystrophy denouncing the Jerry Lewis Telethon as an "Annual Ritual of Shame" and chanting "Power, not pity" before the news cameras? ("MD Telethon," 1991). To date, none of these constituencies has engaged in epistemological interventions that approach, in their depth or extent, AIDS treatment activists' critiques of the methodology of clinical trials. But Bernadine Healy, then the director of the NIH, got it right in 1992 when she told a reporter: "The AIDS activists have led the way. . . . [They] have created a template for all activist groups looking for a cure" (Gladwell, 1992).

Breast cancer activism is an intriguing instance of this new wave, because the links to AIDS activism have been so explicit and so readily acknowledged. In 1991, more than 180 U.S. advocacy groups came together to form the National Breast Cancer Coalition. "They say they've had it with politicians and physicians and scientists who 'there, there' them with studies and statistics and treatments that suggest the disease is under control," read a prominent account in the *New York Times Sunday Magazine* (Ferraro, 1993, p. 26). In its first year of operation, the coalition convinced Congress to step up funding for breast cancer research by $43 million, an increase of almost 50 percent. "The next year, armed with data from a seminar they financed, the women asked for, wheedled, negotiated and won a whopping $300 million more" (Ferraro, 1993, p. 27). The debt to AIDS activism was widely noted by activists and commentators alike. "They showed us how to get through to the Government," said a Bay Area breast cancer patient and organizer: "They took on an archaic system and turned it around while we have been quietly dying." Another activist described how she met with the staff of *AIDS Treatment News* to learn the ropes of the drug development and regulatory systems (Gross, 1991).

Of course, it would be rash to assume that AIDS activism has created an automatic receptiveness on the part of scientists or doctors to health movements of this sort, and that the next round of activists can simply step up to the counter and claim their rewards. A more likely scenario is that AIDS activism will usher in a new wave of democratization struggles in the biomedical sciences and health care—struggles that may be just as hard fought as those of the past decade. It is worth remembering, too, how difficult this sort of activism is to sustain: Organizing a social movement is arduous enough, without having to learn oncology in your spare time.

Complications

Other qualifications to this story deserve notice. Certainly it should be clear that such activism, no matter how broad-ranging it becomes, is unlikely to bring about the thorough transformation of the knowledge-based hierarchies that structure the society we live in. In fact, my analysis suggests a profound tension built into AIDS treatment activists' own project of democratizing expertise. On the one hand, by pursuing an educational strategy to disseminate AIDS information widely, activists have promoted the development of broad-based knowledge-empowerment at the grassroots. On the other hand, as treatment activist leaders have become quasi-experts, they have tended to replicate the expert/lay divide within the movement itself: a small core of activists became insiders who "knew their stuff"; others were left outside to man the barricades. Furthermore, as many of the treatment activists moved "inward," took their seat at the table and became sensitized to the logic of biomedical research, their conceptions of scientific methods sometimes turned in more conventional directions.

"I've seen a lot of treatment activists get seduced by the power, get seduced by the knowledge, and end up making very conservative arguments," contends Michelle Roland (1993), formerly active with ACT UP in San Francisco: "They understand . . . the methodology, they can make intelligent arguments, and it's like, 'Wait a minute . . . okay, you're smart. We accept that. But what's your role?'" Ironically, insofar as activists start thinking like scientists and not like patients, the grounding for their unique contributions to the science of clinical trials may be in jeopardy of erosion. Researcher John Phair (1994) notes that activists "have given us tremendous insight into the feasibility of certain studies," but adds that "some of the activists have gotten very sophisticated and then forget that the idea might not sell" to the community of patients.

Can one be both activist and scientist? Is the notion of a "lay expert" a contradiction in terms? I do not think there are any simple answers here. But arguably, it was not possible for the key treatment activists to become authorities on clinical trials and sit on the ACTG committees, without, in some sense, growing closer to the worldview of the researchers—and without moving a bit away from their fellow activists engaged in other pursuits. Furthermore, the new hierarchies of expertise that have emerged within the ranks of the activists have, to a certain, predictable degree, superimposed themselves on the bedrock of other dimensions of social inequality, including racial, gender, and class differences among activists. And this has led to sharp tensions and outright splits within several activist organizations (Epstein, 1996, pp. 290–294; Epstein, 1997b; Vollmer, 1990; DeRanleau, 1990).

These questions of identity and strategy among knowledge-empowered social movements deserve extended attention. Here I intend only to suggest two implications of my analysis. First, the AIDS activist project of reconfiguring the knowledge-making practices of biomedicine has been executed in ways that are tentative, partial, and shot through with some powerful contradictions. Second, it proves to be not enough to ask what impact AIDS activists have had on the conduct of biomedical research. In addition, we need to ask the reciprocal question: What impact does the encounter with science have on the social movement? How does the "expertification" or "scientization" of activism affect the goals and tactics of a social movement, as well as its collective identity (Epstein, 1997b)? Without doubt, the reciprocal relation between AIDS activists and AIDS researchers was equally transformative in both directions.

Let me now raise a final worry about the democratization of science: the obvious risk that lay participation will interfere with the good conduct of science and indeed delay the goals that all want to see achieved. What is to prevent real harm from being done? In this regard, it is important to note that the process of activist intervention in biomedicine is not without some painful ironies. On the one hand, the enterprise appears significantly driven by the dictates of expediency and dire need—"I'm dying so give me the drug now!"; on the other hand, the core treatment activists have increasingly become believers in science (however understood), and desperately want clinical trials to generate usable knowledge that can guide medical practice. As David Barr of the Treatment Action Group (a spin-off of ACT UP/New York) put it: "My doctors and I make decisions in the dark with every pill I put in my mouth" (Cotton, 1991, p. 1362)—and this is not an easy way to live.

Insofar as activists want clinical trials to succeed, they must wrestle with the consequences of their own interventions. Do such interventions enhance activists' capacity to push clinical research in the directions they choose? Or do activists and researchers alike become subject to the unintended effects of their actions, trapped within an evolving system whose trajectory no one really controls? Here is a sort of worst-case scenario of the spiraling consequences of community-based interventions in the construction of belief in antiviral drugs—a caricature sketch, to be sure, but one that combines elements from a number of cases in the late 1980s and early 1990s. Drug X performs well in preliminary studies, and an NIH official is quoted as saying that X is a promising drug. The grassroots treatment publications write that X is the up-and-coming thing; soon everyone in the community wants access to X, and activists are demanding large, rapid trials to study it. Everyone wants to be in the trial, because they believe that X will help them; but researchers want to conduct the trial in order to determine whether X has any efficacy. Those who cannot get into the trial demand expanded access, while others begin importing X from other countries or manufacturing it in clandestine laboratories. As X becomes more prevalent and emerges as the de facto standard of care, physicians begin to suggest that patients get hold of it however they can. Meanwhile, participants in the clinical trial of X who fear they are receiving a placebo mix and match their pills with other participants. When the trial's investigators report potential treatment benefits, activists push for accelerated approval of X, leaving the final determination of X's efficacy to postmarketing studies. But who then wants to sign up for those studies, when everyone now believes that the drug works, since, after all, the FDA has licensed it and any doctor can prescribe it?

This is a scary scenario, but it must be pointed out that, in recent years, activists themselves have sought to control this troubling escalation and to extricate themselves from what they rightly call the "hype cycle." As Mark Harrington (1993, p. 7) wrote in late 1993: "One disturbing but inevitable result of the urgency engendered by the AIDS crisis is that both researchers and community members tend to invest preliminary trials with more significance than they can possibly bear." To the extent that activists can develop a critique of this phenomenon of expecting too much from research, and to the extent that they can communicate the *relative uncertainty* of clinical trials to the broader public of HIV-infected persons, it may be possible to imagine a clinical research process that more fully reflects the interests of those who are most in need of answers.[6]

Notes

1. This analysis of AIDS treatment activists derives from a larger research project concerned with studying the conduct of science in the AIDS epidemic and the role of laypeople, and particularly activists, in the transformation of biomedical knowledge-making practices (see especially Epstein, 1996). Much of the text of this article has appeared previously (see Epstein, 1995, 1996, 1997a, 1997b). This account is based on interviews conducted with AIDS activists, AIDS researchers, and government health officials at the National Institutes of Health and the Food and Drug Administration; as well as on analyses of the accounts, claims-making, and framing of issues presented in scientific and medical journals, the mass media, the gay and lesbian press, activist publications, activist documents, and government documents.

2. On the politics of public participation in science and medicine, see, for example, Balogh (1991); Blume, Bunders, Leydesdorff, and Whitley (1987); Brown (1992); Cozzens and Woodhouse (1995); Di Chiro (1992); Indyk and Rier (1993); Irwin and Wynne (1996); Kleinman (1995); Martin (1980); Moore (1996); Nelkin (1975); Petersen (1984); Rycroft (1991); White (1993); and Wynne (1992).

3. My conception of credibility borrows from scholarship in science studies that includes Barnes (1985); Barnes and Edge (1982); Cozzens (1990); Latour and Woolgar (1986); Shapin (1994); Shapin and Schaffer (1985); Star (1989); and Williams and Law (1980).

4. To borrow Bruno Latour's (1987, p. 132) term, activists constructed themselves as an "obligatory passage point" standing between researchers and the trials they sought to conduct. Of course, activists also needed the researchers to conduct the trials, so the relationship is best seen as symbiotic. See also Crowley (1991).

5. Here I mean to go beyond the argument, now routinely heard, that AIDS has forever changed conceptions of the doctor-patient relationship. That may be true, although probably the old-fashioned model of the omnipotent physician and the dependent patient was already on the way out.

6. This perspective brings activists into alignment with sociologists of scientific knowledge who advocate that public understanding of science can be improved if the public acquires a greater appreciation of the high degree of uncertainty in science (see Collins & Pinch, 1993).

Bibliography

Adam, B. D. (1987). *The rise of a gay and lesbian movement.* Boston: Twayne Publishers.

Altman, D. (1982). *The homosexualization of America.* Boston: Beacon Press.

Altman, D. (1986). *AIDS in the mind of America.* Garden City, NJ: Anchor Press/Doubleday.

Altman, D. (1994). *Power and community: Organizational and cultural responses to AIDS.* London: Taylor & Francis.

Anonymous. (1991). *ACT UP/New York capsule history.* New York: AIDS Coalition to Unleash Power.

Antiviral Drugs Advisory Committee of the U.S. Food and Drug Administration. (1991, February 13–14). Meeting transcript. Bethesda, MD: Food and Drug Administration.

Arno, P. S., & Feiden, K. L. (1992). *Against the odds: The story of AIDS drug development, politics and profits.* New York: HarperCollins.

Balogh, B. (1991). *Chain reaction: Expert debate and public participation in American commercial nuclear power, 1945–1975.* Cambridge, England: Cambridge University Press.

Barinaga, M. (1992). Furor at lyme disease conference. *Science, 256,* 1384–1385.

Barnes, B. (1985). *About science.* Oxford: Basil Blackwell.

Barnes, B., & Edge, D. (1982). Science as expertise. In *Science in context: Readings in the sociology of science,* B. Barnes & D. Edge (Eds.), (pp. 233–249). Cambridge, MA: MIT Press.

Bayer, R. (1981). *Homosexuality and American psychiatry: The politics of diagnosis.* New York: Basic Books.

Bayer, R. (1985). AIDS and the gay movement: Between the specter and the promise of medicine. *Social Research, 52,* 581–606.

Blume, S., Bunders, J., Leydesdorff, L., & Whitley, R. (Eds.). (1987). *The social direction of the public sciences.* Dordrecht, Holland: D. Reidel.

Boston Women's Health Book Collective. (1973). *Our bodies, ourselves: A book by and for women.* New York: Simon & Schuster.

Bourdieu, P. (1990). *The logic of practice.* Stanford, CA: Stanford University Press.

Brown, P. (1992). Popular epidemiology and toxic waste contamination: Lay and professional ways of knowing. *Journal of Health and Social Behavior, 33,* 267–281.

Bull, C. (1988, October 16–22). Seizing control of the FDA. *Gay Community News, 1,* 3.

Cohen, C. J. (1993). Power, resistance and the construction of crisis: Marginalized communities respond to AIDS. Unpublished doctoral dissertation, University of Michigan.

Collins, H,, & Pinch, T. (1993). *The golem: What everyone should know about science.* Cambridge, England: Cambridge University Press.

Corea, G. (1992). *The invisible epidemic: The story of women and AIDS.* New York: HarperCollins.

Cotton, P. (1991, March 20). HIV surrogate markers weighed. *Journal of the American Medical Association* 265:11:1357, 1361, 1362.

Cozzens, S. E. (1990). Autonomy and power in science. In S. E. Cozzens & T. F. Gieryn (Eds.), *Theories of science in society,* (pp. 164–184). Bloomington: Indiana University Press.

Cozzens, S. E., & Woodhouse, E. J. (1995). Science, government, and the politics of knowledge. In S. Jasanoff, G. Markle, J. C. Petersen, & T. Pinch (Eds.), *Handbook of science and technology studies,* (p. 533–553. Thousand Oaks, CA: Sage.

Crimp, D., & Rolston, A. (1990). *AIDS demographics.* Seattle: Bay Press.

Crowley, W. F. P. III. (1991). Gaining access: The politics of AIDS clinical drug trials in Boston. Unpublished undergraduate thesis, Harvard College.

DeRanleau, M. (1990, October 1). How the "conscience of an epidemic" unraveled. *San Francisco Examiner,* A-15.

Di Chiro, G. (1992). Defining environmental justice: Women's voices and grassroots politics. *Socialist Review 22,* 93–130.

Elbaz, G. (1992). The sociology of AIDS activism, the case of ACT UP/New York, 1987–1992. Unpublished doctoral dissertation, City University of New York.

Emke, I. (1993). Medical authority and its discontents: The case of organized non-compliance. *Critical Sociology, 19,* 57–80.

Epstein, S. (1995). The construction of lay expertise: AIDS activism and the forging of credibility in the reform of clinical trials. *Science, Technology, & Human Values, 20,* 408–437.

Epstein, S. (1996). *Impure science: AIDS, activism, and the politics of knowledge.* Berkeley: University of California Press.

Epstein, S. (1997a). Activism, drug regulation, and the politics of therapeutic evaluation in the AIDS era: A case study of ddC and the "surrogate markers" debate. *Social Studies of Science, 27,* 691–726.

Epstein, S. (1997b). AIDS activism and the retreat from the genocide frame. *Social Identities, 3,* 415–438.

Fauci, A. (1994, October 31). Interview by author. Bethesda, MD.

FDA allows AIDS patients to import banned drugs. (1988, July 24). *The Los Angeles Times,* p. 18.

Fee, E. (Ed.). (1982). *Women and health: The politics of sex in medicine.* Farmingdale, NY: Baywood.

Ferraro, S. (1993, August 15). The anguished politics of breast cancer. *The New York Times Sunday Magazine,* pp. 25–27, 58–62.

Gamson, J. (1989). Silence, death, and the invisible enemy: AIDS activism and social movement "newness." *Social Problems, 36,* 351–365.

Garrison, J. (1989, September 5). AIDS activists being heard. *The San Francisco Examiner,* pp. A-1, A-8.

Geltmaker, T. (1992). The queer nation acts up: Health care, politics, and sexual diversity in the county of angels. *Society and Space, 10,* 609–650.

Gladwell, M. (1992, October 15). Beyond HIV: The legacies of health activism. *The Washington Post,* p. A-29.

Gross, J. (1991, January 7). Turning disease into political cause: First AIDS, and now breast cancer. *The New York Times,* p. A-12.

Harrington, M. (1993). *The crisis in clinical AIDS research.* New York: Treatment Action Group.

Harrington, M. (1994, April 29). Interview by author. New York City.

Hilts, P. J. (1990, May 22). 82 held in protest on pace of AIDS research. *The New York Times,* p. C-2.

Indyk, D., & Rier, D. (1993). Grassroots AIDS knowledge: Implications for the boundaries of science and collective action. *Knowledge: Creation, diffusion, utilization, 15,* 3–43.

Irwin, A., & Wynne, B. (1996). *Misunderstanding science? The public reconstruction of science and technology.* Cambridge, England: Cambridge University Press.

James, J. S. (1986, May 9). What's wrong with AIDS treatment research? *AIDS Treatment News.*

Jasper, J. M., & Nelkin, D. (1992). *The animal rights crusade: The growth of a moral protest.* New York: Free Press.

Jonsen, A. R.,& Stryker, J. (Eds.). (1993). *The social impact of AIDS in the United States.* Washington, DC: National Academy Press.

Kingston, T. (1991, November 7). The "white rats" rebel: Chronic fatigue patients sue drug manufacturer for breaking contract to supply promising CFIDS drug. *The San Francisco Bay Times,* pp. 8, 44.

Kleinman, D. L. (1995). *Politics on the endless frontier: Postwar research policy in the United States.* Durham, NC: Duke University Press.

Kroll-Smith, S., & Floyd, H. H. (1997). *Bodies in protest: Environmental illness and the struggle over medical knowledge.* New York: New York University Press.

Kwitny, J. (1992). *Acceptable risks.* New York: Poseidon Press.

Latour, B. (1987). *Science in action: How to follow scientists and engineers through society.* Cambridge, MA: Harvard University Press.

Latour, B., & Woolgar, S. (1986). *Laboratory life: The construction of scientific facts.* Princeton, NJ: Princeton University Press.

Lo, C. Y. H. (1992). Communities of challengers in social movement theory. In A. D. Morris & C. M. Mueller (Eds.), *Frontiers in social movement theory,* (pp. 224–247). New Haven: Yale University Press.

Martin, B. (1980). The goal of self-managed science: Implications for action. *Radical Science Journal, 10,* 3–16.

MD Telethon Boycott Urged. (1991, September 2). *The San Francisco Examiner,* B-1.

Merigan, T. C. (1990). Sounding board: You *can* teach an old dog new tricks: How AIDS trials are pioneering new strategies. *New England Journal of Medicine, 323,* 1341–1343.

Meyer, D. S., & Whittier, N. (1994). Social movement spillover. *Social Problems, 41,* 277–298.

Moore, K. (1996). Organizing integrity: American science and the creation of public interest organizations, 1955–1975. *American Journal of Sociology, 101,* 1592–1627.

Nelkin, D. (1975). The political impact of technical expertise. *Social Studies of Science, 5,* 35–54.

Patton, C. (1990). *Inventing AIDS.* New York: Routledge.

Petersen, J. C. (Ed.). (1984). *Citizen participation in science policy.* Amherst: University of Massachusetts Press.

Phair, J. (1994, November 15). Interview by author. Chicago.

PWA Coalition. (1988). Founding statement of people with AIDS/ARC. In D. Crimp (Ed.), *AIDS: Cultural analysis, cultural activism* (pp. 148–149). Cambridge, MA: MIT Press.

Quimby, E., & Friedman, S. R. (1989). Dynamics of black mobilization against AIDS in New York City. *Social Problems, 36,* 403–415.

Richman, D. (1994). Interview by author. San Diego, June 1.

Roland, M. 1993. Interview by author. Davis, CA, December 18.

Ruzek, S. B. (1978). *Feminist alternatives to medical control.* New York: Praeger.

Rycroft, R. W. (1991). Environmentalism and science: Politics and the pursuit of knowledge. *Knowledge: Creation, Diffusion, Utilization, 13,* 150–169.

Shapin, S. (1994). *A social history of truth: Civility and science in seventeenth-century England.* Chicago: University of Chicago Press.

Shapin, S., & Schaffer, S. (1985). *Leviathan and the air-pump: Hobbes, Boyle, and the experimental life.* Princeton, NJ: Princeton University Press.

Star, S. L. (1989). *Regions of the mind: Brain research and the quest for scientific certainty.* Stanford, CA: Stanford University Press.

Treichler, P. A. (1991). How to have theory in an epidemic: The evolution of AIDS treatment activism. In C. Penley & A. Ross (Eds.), *Technoculture* (pp. 57–106). Minneapolis and Oxford: University of Minnesota Press.

Vollmer, T. (1990, September 20). ACT-UP/SF splits in two over consensus, focus. *San Francisco Sentinel,* 1.

Wachter, R. M. (1991). *The fragile coalition: Scientists, activists, and AIDS.* New York: St. Martin's.

Weber, M. (1978). *Economy and society,* vol. 1. G. Roth & C. Wittich (Eds.). Berkeley: University of California Press.

White, S. (1993). Scientists and the environmental movement. *Chain Reaction,* *68,* 31–33.

Williams, R., & Law, J. (1980). Beyond the bounds of credibility. *Fundamenta Scientiae, 1,* 295–315.

Winnow, J. (1992). Lesbians evolving health care: Cancer and AIDS. *Feminist Review, 41,* 68–77.

Wynne, B. (1992). Misunderstood misunderstanding: Social identities and public uptake of science. *Public Understanding of Science, 1,* 281–304.

2

Town Meetings on Technology

Consensus Conferences as Democratic Participation

RICHARD E. SCLOVE

In a democracy, it normally goes without saying that policy decisions affecting all citizens should be made democratically. Science and technology policies loom as grand exceptions to this rule. They certainly affect all citizens profoundly: the world is continuously remade with advances in telecommunications, computers, materials science, weaponry, biotechnology, home appliances, energy production, air and ground transportation, and environmental and medical understanding. Yet science and technology policies are customarily framed by representatives of just three groups: business, the military, and universities (Sclove, 1998; Dickson, 1984/1988). These are the groups invited to testify at congressional hearings, serve on government advisory panels, and prepare influential policy studies.

According to conventional wisdom, the reason for this state of affairs is that nonexperts are ill-equipped to comment on complex technical matters. It seems unimaginable that citizens who can't even program their VCRs could ever contribute constructively to complex scientific and industrial issues. But a wide range of emerging social innovations flatly contradict traditional attitudes. Among them is the consensus conference.[1] This organizational form was pioneered during the late 1980s by the Danish Board of Technology, a parliamentary agency charged with assessing technologies, and its successful utilization in Denmark has led to recent experiments with the practice elsewhere in Europe and in Japan. In 1997 the nonprofit Loka Institute, which I direct, initiated and co-organized a pilot consensus conference in the United States.

Consensus conferences are intended to stimulate broad and intelligent social debate on technological issues. Not only are laypeople elevated to positions of preeminence, but a carefully planned program of reading and discussion, culminating in a forum open to the public, ensures that they become well-informed prior to rendering judgment.

Both the forum and the subsequent judgment, written up in a formal report, become a focus of intense national attention in Denmark—usually at a time when the issue at hand is due to come before Parliament. Though consensus conferences are hardly meant to dictate public policy—indeed, their judgments are nonbinding—they do give legislators some sense of where the people who elected them might stand on important questions. They can also help industry steer clear of new products or processes that are likely to spark public opposition.

Since 1987, the Danish Board of Technology has organized about twenty consensus conferences on topics ranging from genetic engineering to educational technology, food irradiation, air pollution, human infertility, sustainable agriculture, telecommuting, and the future of private automobiles.[2] Ironically, the popularity of the process began to grow and diffuse internationally just as the U.S. Congress was eliminating its Office of Technology Assessment (OTA) in 1995 (Bimber & Guston, 1997). The establishment of the OTA in 1972 helped motivate Europeans to develop their own technology assessment agencies. But the truth is that when the OTA faced the chopping block, those rallying to its defense were primarily a small cadre of professional policy analysts or other experts who had themselves participated in OTA studies—hardly a sizable cross-section of the American public. By contrast, a consensus conference format, which engages a much wider range of people, holds the potential to build a broader constituency familiar with and supportive of technology assessment. Building such a constituency, and the widespread implementation of processes like the consensus conference, will be crucial to limiting the negative and often unintended consequences that can result when technologies are deployed without widespread social consideration. Moreover by increasing citizen involvement in civic life, consensus conferences can play a role in combating cynicism and rebuilding a vibrant democratic culture in the United States.[3]

Framing the Issues

To organize a consensus conference, the Danish Board of Technology first selects a salient topic—one that is of social concern, pertinent to upcoming parliamentary deliberations, and complex, requiring judgment on such diverse matters as ethics, disputed scientific claims, and government policy. The board has also found that topics suited to the consensus conference format should be intermediate in scope—broader than assessing the toxicity of a single chemical for instance, but narrower than trying to formulate a comprehensive national environmental strategy. The board then chooses a well-balanced steering committee to oversee the organization of the conference; a typical committee might include an academic scientist, an industry researcher, a trade unionist, a representative of a public-interest group, and a project manager from the board's own professional staff.

With the topic in hand and the steering committee on deck, the board advertises in local newspapers throughout Denmark for volunteer lay participants. Candidates must send in a one-page letter describing their backgrounds and their reasons for wanting to participate. From the 100 to 200 replies that it receives, the board chooses a panel of about 15 people who roughly represent the demographic breadth of the Danish population and who lack significant prior knowledge of, or a specific material interest in, the topic. Groups include homemakers, office and factory workers, and garbage collectors as well as university-educated professionals. They are not, however, intended to comprise a random scientific sample of the Danish population. After all, each panelist is literate and motivated enough to have responded in writing to a newspaper advertisement.

At the outset of a first preparatory weekend meeting, the lay group, with the help of a skilled facilitator, discusses an expert background paper commissioned by the board and screened by the steering committee that maps the political terrain surrounding the chosen topic. The lay group next begins formulating questions to be addressed during the public forum. Based on the lay panel's questions, the board goes on to assemble an expert panel that includes not only credentialed scientific and technical experts but also experts in ethics or social science and knowledgeable representatives of stakeholder groups such as trade unions, industry, and environmental organizations.

The lay group then meets for a second preparatory weekend, during which members, again with the facilitator's help, discuss more background readings provided by the steering committee, refine their questions, and if they want, suggest additions to or deletions from the expert panel. Afterward, the board finalizes selection of the expert panel and asks its members to prepare succinct oral and written responses to the lay group's questions, expressing themselves in language that laypeople will understand.

The concluding public forum, normally a four-day event chaired by the facilitator who presided over the preparatory weekends, brings the lay and expert panels together and draws the media, members of Parliament, and interested Danish citizens. On the first day each expert speaks for 20 to 30 minutes and then addresses follow-up questions from the lay panel and, if time allows, the audience. Afterward, the lay group retires to discuss what it heard. On the second day the lay group publicly cross-examines the expert panel in order to fill in gaps and probe further into areas of disagreement.

Once cross-examination has been completed, the experts and stake-holder representatives are politely dismissed. The remainder of that day and on through the third, the lay group prepares its report, summarizing the issues on which it could reach consensus and identifying any remaining points of disagreement. The board provides secretarial and editing assistance, but the lay panel retains full control over the report's content. On the fourth and final day, the expert group has a brief opportunity to correct outright misstatements of their testimony in the report, but not otherwise to comment on the document's substance. Directly afterward, the lay group presents its report at a national press conference.[4]

Lay panel reports are typically 15 to 30 pages long, clearly reasoned, and nuanced in judgment. The report from the 1992 Danish conference on genetically engineered animals is a case in point, showing a perspective that is neither pro- nor antitechnology in any general sense. The panel expressed concern that patenting animals could deepen the risk of their being treated purely as objects. Members also feared that objectification of animals could be a step down a slippery slope toward objectification of people. Regarding the possible ecological consequences of releasing genetically altered animals into the wild, they noted that such animals could dominate or out-compete wild species or transfer unwanted characteristics to them. However, the group saw no appreciable ecological hazard in releasing genetically engineered cows or other large domestic animals into fenced fields, and endorsed deep-freezing animal sperm cells and eggs to help preserve biodiversity (Consensus Conference, 1998).

Portions of the lay reports can be incisive and impassioned as well, especially in comparison with the circumspection and the dry language that is conventional in expert policy analyses. Having noted that the "idea of genetic normalcy, once far-fetched, is drawing close with the development of a full genetic map," a 1988 OTA study of human genome research concluded blandly that "concepts of what is normal will always be influenced by cultural variations" (OTA, 1988, p. 85).[5] In contrast, a 1989 Danish consensus panel on the same subject recalled the "frightening" eugenic programs of the 1930s and worried that "the possibility of diagnosing fetuses earlier and earlier in pregnancy in order to find 'genetic defects' creates the risk of an unacceptable perception of man—a perception according to which we aspire to be perfect" (Consensus Conference, 1989, pp. 6, 14, 17). The lay group went on to appeal for further popular debate on the concept of normalcy. Fearing that parents might one day seek abortions on learning that a fetus was, say, color blind or left-handed 14 of the panel's 15 members also requested legislation that would make fetal screening for such conditions illegal under most circumstances (Consensus Conference, 1989, pp. 17–18, 26).

This central concern with social issues becomes much more likely when expert testimony is integrated with everyday citizen perspectives. For instance, while the executive summary of the OTA study on human genome research states that "the core issue" is how to divide up resources so that genome research is balanced against other kinds of biomedical and biological research (OTA, 1988, p. 10), the Danish consensus conference report, prepared by people whose lives are not intimately bound up in the funding dramas of university and national laboratories, opens with a succinct statement of social concerns, ethical judgments, and political recommendations. And these perspectives are integrated into virtually every succeeding page, whereas the OTA study discusses ethics only in a single discrete chapter on the subject. The Danish consensus conference report concludes with a call for more school instruction in "subjects such as biology, religion, philosophy, and social science"; better popular dissemination of "immediately understandable" information about genetics; and vigorous government efforts to promote the broadest popular discussion of "technological and ethical issues" (Consensus Conference, 1989, pp. 28, 29). The corresponding OTA study does not even consider such ideas.

When the Danish lay group did address the matter of how to divide up resources, they differed significantly from the OTA investigators. Rather than focusing solely on balancing different kinds of biomedical and biological research against one another, they supported basic research in genetics but also called for more research on the interplay between environmental factors and genetic inheritance, and more research on the social consequences of science. They challenged the quest for exotic technical fixes for diseases and social problems, pointing out that many proven measures for protecting health and bettering social conditions and work environments are not being applied. Finally, they recommended a more "humanistic and interdisciplinary" national research portfolio that would stimulate a constructive exchange of ideas about research repercussions and permit "the soul to come along" (Consensus Conference, 1989, pp. 7, 17–25).

Not that consensus conferences are better than the OTA approach in every possible way. While less accessibly written and less attentive to social considerations, a traditional OTA report provides more technical detail and analytic depth. But OTA-style analysis can, in principle, contribute to the consensus conference process. For example, the 1993 Dutch consensus conference on animal biotechnology used a prior OTA study as a starting point for its own more participatory inquiry.[6]

Timeliness and Responsiveness

Once the panelists have announced their conclusions, the Danish Board of Technology exemplifies its commitment to encouraging informed discussion by publicizing them through local debates, leaflets, and videos. In the case of biotechnology, the board subsidized more than 600 local debate meetings. The board also works to ensure that people are primed for this whirlwind of postconference activity. For example, the final four-day public forums are held in the Parliament building, where they are easily accessible to members of Parliament and the press.

Nor is it any accident that the topics addressed in consensus conferences are so often of parliamentary concern when the panelists issue their findings. The board has developed the ability to organize a conference on six months notice or less largely for the purpose of attaining that goal. This timeliness represents yet another advantage over the way technology assessment has been handled in the United States: relying mostly on lengthy analysis and reviews by experts and interest groups, the OTA required, on average, two years to produce a published report on a topic assigned by Congress. In fact, one complaint leveled by the congressional Republicans who argued for eliminating the agency was that the process it employed was mismatched to legislative timetables. On learning about consensus conferences and their relatively swift pace, U.S. Congressman Robert S. Walker—at the time Republican chair of the House Science Committee—told a March 1995 public forum that if such a process can "cut down the time frame and give us useful information, that would be something we would be very interested in" (Walker, 1995).

The Board of Technology's efforts do seem to be enhancing public awareness of issues in science and technology. A 1991 study by the European Commission discovered that Danish citizens were better informed about biotechnology, a subject that several consensus conferences had addressed, than were the citizens of other European countries, and that Danes were relatively accepting of their nation's biotechnology policies as well (INRA 1995). Public opinion surveys performed in 1995 reveal that, as a cumulative result of organizing successive consensus conferences over the course of a decade, approximately 35 percent of the Danish public is now acquainted with the process (Joss, 2000). Significantly too, Dr. Simon Joss of the Centre for the Study of Democracy in London, who has conducted interviews on consensus conferences with Danish members of Parliament, has found the legislators to be generally appreciative of the process—indeed, to the point where several eagerly pulled down conference reports kept at hand on their office shelves.[7]

And although consensus conferences are not intended to have a *direct* impact on public policy, they do in some cases. For instance, conferences that were held in the late 1980s influenced the Danish Parliament to pass legislation limiting the use of genetic screening in hiring and insurance decisions, to exclude genetically modified animals form the government's initial biotechnology research and development program, and to prohibit food irradiation for everything except dry spices (Klüver, 1995, p. 44). Manufacturers are taking heed of the reports that emerge from consensus conferences as well. According to Dr. Tarja Cronberg, in a report issued by the Technical University of Denmark, Danish industry originally resisted even the idea of establishing the Board of Technology but has since had a change of heart (Cronberg, n.d.). The reasons are illuminating.

In conventional politics of technology, the public's first opportunity to react to an innovation can occur years or even decades after crucial decisions about the form that innovation will take have already been made. In such a situation, the only feasible choice is between pushing the technology forward or bringing everything to a halt. And no one really wins: pushing the technology forward risks leaving opponents bitterly disillusioned, whereas bringing everything to a halt can jeopardize jobs and enormous investments of developmental money, time, and talent. The mass movements of the 1970s and 1980s that more or less derailed nuclear power are a clear example of the phenomenon.

By contrast, early public involvement and publicity—of the sort that a consensus conference permits—can facilitate more flexible, socially responsive research and design modifications all along the way. This holds the potential for a fairer, less adversarial, and more economical path of technological evolution (Sclove, 1995, pp. 183–184). A representative of the Danish Council of Industry related that corporations have benefitted from their nation's participatory approach to technology assessment because "product developers have worked in a more critical environment, thus being able to forecast some of the negative reactions and improve their products in the early phase" (quoted in Cronberg, n.d., p. 11).

For example, directors of Novo Nordisk, a large Danish biotechnology company, reevaluated their research and development strategies after a 1992 consensus conference report deplored the design of animals suited to the rigors of existing agricultural systems but endorsed the use of genetic engineering to help treat incurable diseases.[8]

A First: Boston's Consensus Conference

In 1997, residents of the greater Boston area made history, participating in the first ever consensus conference in the United States. The conference, dubbed a "Citizens' Panel," was initiated by the Loka Institute.[9] The principle organizers included Loka, the staff and students of the Education for Public Inquiry and International Citizenship (EPIIC) Program at Tufts University, the Massachusetts Foundations for the Humanities, and MIT's *Technology Review* magazine. Other supporting or assisting organizations included the

Jefferson Center based in Minneapolis, the University of Massachusetts Extension Program and the School of Behavioral and Social Sciences at UMass-Amherst, the National Science Foundation, the John D. and Catherine T. MacArthur Foundation, and the Benton Foundation. The topic was telecommunications and the future of democracy, and the 15-member citizens' panel issued a call for protecting First Amendment rights and personal privacy on the Internet, mandating community involvement in telecommunications policy-making, and returning a percentage of high-tech corporate earnings to communities and nonprofit organizations.

Selected by random telephone calling and supplementary targeted recruitment to be broadly representative of wider Boston's population, the Citizens' Panel members included an auto mechanic, the business manager of a high-tech firm, a retired teacher/farmer/nurse, and an industrial engineer. There was also an arts administrator, a 1996 innercity high school graduate, a consultant, an unemployed social worker, a writer/actress, and a homeless shelter resident. Eight of the panelists were women; 7 were men. Five of the 15 were people of color, and their life stages ranged from teenager through elder.

During February and March, the panelists met together over two weekends to discuss background readings and introductory briefings on telecommunication issues. Then on April second and third, all fifteen panelists braved a city-crippling, two-foot snow storm to hear ten hours of expert testimony from computer specialists, government officials, and business executives. Among those giving testimony were the president of New England Cable News, an official of Lotus Development Corporation, the Congressional Liaison to the United States Department of Commerce who helped draft the 1996 Telecommunications Reform Act, a school superintendent, and representatives of public-interest groups.

After deliberating and drafting their own report, the lay panel reconvened on the morning of April fourth to announce their findings at a press conference organized at Tufts University. A WCVB/CNN television crew was on hand to record their performance. Lay panelists, who had been presented with expert testimony that included a string of vigorous business perspectives, came out in favor of a judicious but far-reaching public-interest agenda—a more ambitious program than anything embodied in the 1996 Telecommunications Reform Act. Their report urged governments to establish more forums for citizen participation in policy issues, even on highly technical matters like telecommunications. And the document argues that:

> Business interests, profit motives and market forces too often dictate public policy to the exclusion of the interests of the people (an example of which is the 1996 Telecommunications Act). The new

technology creates an even greater risk of the abuse of power. (Consensus Statement, 1997)

The timing of the panel's report was strategic, because this is a watershed period in U.S. telecommunications policy-making. For instance, at the time of the Boston consensus conference the Federal Communications Commission (FCC) was—as required by the Telecommunications Reform Act—working on recommendations for implementing universal Internet access, and had just completed the auctioning of digital audio broadcast licenses. The Supreme Court had just heard arguments for and against the free speech-inhibiting Communications Decency Act. And the Clinton administration had recently appointed an advisory committee on the public-interest obligations of digital broadcasters and was in the midst of its initiative to wire all schools to the Internet.

The report of the Citizen Panel included a number of specific recommendations. Among them were:

- Protecting privacy and First Amendment rights on the Internet and affirming a standard of personal responsibility in using on-line material;
- Establishing volunteer citizens' groups at the local level to address appropriate restriction of access to certain (e.g., pornographic) Internet sites at public libraries, schools, and community centers;
- Encouraging businesses to return a percentage of their profits to the local communities they serve;
- Legally prohibiting the use of private individual data without prior notification and approval;
- Making Internet-connected school computers available to the general public for lifelong learning outside school hours; and
- Extending "universal access" beyond infrastructural development to encompass "universal service," in order to insure that the general public has both facilities and the opportunity to log on. (Consensus Statement, 1997)

This first ever U.S. consensus conference was important for several reasons. It constituted the first systematic attempt in the United States to solicit informed input from ordinary citizens—including six who had never previously used the Internet, half of whom had also never used a computer—on the complexities of current telecommunications policy. Telecommunications aside, this was also the first time in modern U.S. history that a diverse group of everyday citizens—none previously expert on the policy issues under discussion, none a representative from an organization with a direct stake (not even from a public-interest group)—gathered to learn and deliberate on a scientific or technological topic of this breadth or complexity.[10] What is more, the topic investigated by panelists—"Telecommunications and the Future of Democracy"—was broader than topics that have been addressed in Danish and other European consensus conferences, suggesting that the methodology may have wider applicability than previously understood.

Some Observations

I spent three years working toward this event. During that time, innumerable doubters contended that a participatory process invented in Denmark (where, as the stereotype would have it, "everyone is white, tall, blond, educated, affluent, and civic-minded") could never work in the United States. Americans are too apathetic, too ill-educated, and too different from one another. For instance, a project director at the OTA (when it still existed) insisted that the agency had tried repeatedly to involve ordinary citizens in its report review process, but that citizens simply refused to participate.

Our consensus conference proved the skeptics wrong. On a first try we were able to assemble a lay panel more diverse than any gathered as of that date in Europe.[11] All 15 members attended both background weekends and the final forum. Watching the lay panelists both listen to and interrogate expert witnesses, I saw no yawns, no wandering eyes, no fussing with hair. The panelists listened closely and asked one astute question after another. Indeed, because the background weekends effectively brought lay panelists up-to-speed on telecommunications issues, their questions were sometimes more technical than the experts' testimony.

We were also able to undertake this endeavor on a relative shoe-string. The budget for this pilot program was about $60,000. European consensus conferences have typically cost between $100,000 and $200,000. Some of this expense reflects the fact that European consensus conferences have been nationwide, and consequently organizers had to pay for participants' travel and lodging. A nationwide panel undertaken in the United States would cost somewhat more—probably on the order of $500,000. That is a lot of money, but still trivial compared with the expenditures and social impacts that are at stake in major technology policy decisions—for example, annual U.S. expenditure on R&D, from public and private sources, currently tops $200 billion (National Science Board, 1998, p. A-121). If a small nation like Denmark can routinely afford $100,000 for consensus conferences, I find it incredible that we cannot afford just five times that. We are after all a wealthy nation, and our population is more than 50 times that of Denmark.

If this initial effort was cost-effective, it was also efficiently organized. From the time of assembling the initial institutional partners for the Boston event, it took just five months to do the preliminary planning, assemble the minimal necessary funding, and hire a project manager.[12] From the date of hiring our manager until the concluding public forum took another five months. This was swift—about the same time required by the Danish Board of Technology, and that body has a great deal of experience, having organized many consensus conferences over the past decade. By comparison, the first British consensus conference, which concluded in November of 1994, took about a year and a half to organize.[13]

Beyond the substantive successes of the conference, lay panelists' comments suggest that their work helped built a sense of community and foster feelings of citizenship. One panelist noted that working together "tore down walls." "You don't usually speak to such diverse other people in daily life, like when you're riding on a bus," continued this participant. At the final press conference, another panelist concluded that, "We need more panels like this, to give us the opportunity to learn and to take what we learn back to our communities." Finally, another participant in the Boston consensus conference said:

> I became proud of my sense of citizenship. I began to identify with the people of ancient Athens. . . . There was a wonderful sense of belonging, and of being able to make a difference when a group convenes.

For a pilot project, I think ours was a tremendous success. Nevertheless, there were difficulties and shortcomings, and attentiveness to these could make future consensus conferences in the United States even more successful.[14] First, there was not enough time and staffing to support adequate consultations between the project manager and the project steering committee, a diverse group of knowledgeable stakeholders chosen to help ensure impartiality in the organization. Second, while the expert panel was reasonably well-balanced between academics, industry, government, and public-interest groups, imbalances in outlook did surface. For example, in the case of one subtopic—computers in schools—the lay panelists heard three very similar upbeat presentations by outspoken proponents of computers in education and not a single off-setting critical perspective. Whether the lay panel would have reached different conclusions had they heard a more diverse set of views is difficult to know. However as a rule of thumb, I believe there should be a minimum of three very different expert opinions presented on each contested issue. Such diversity enhances the likelihood of provoking nuanced, reflective discussion on the issue among the lay panelists.

A third difficulty arose from institutional and budget constraints. While the procedural and substantive results from this initial U.S. consensus conference were impressive by many measures, lacking government sponsorship or a budget to pay expert honoraria, we were unable to secure a commitment from many of our expert witnesses to attend for two days. Thus, we had to omit a key component of the Danish consensus conference methodology: the lay panelists' open cross-examination of all the expert witnesses assembled together on a second day. Cross-examination gives the lay panel a chance to play off expert witnesses against one another, and thus to take their own knowledge and judgment to a higher level of integration.[15] Finally, while covered on one local television station, in the *Boston Globe* newspaper, and in *Technology Review* magazine (Flint, 1997; Hackman, 1997), the consensus conference did not receive as much media coverage as an event of this importance deserves. Some of the explanation for this can be attributed to the freak April first Boston snowstorm that closed down Logan airport for two days and forced a number of national reporters to cancel their plans to attend. Beyond this, however, nonpartisan or bipartisan government sponsorship, or a budget to bring in more high profile expert witnesses, would presumably also help improve media attentiveness.

Conclusion

United States science and technology institutions and decision-making processes stand out among many industrialized nations for systematically excluding lay citizen voices. The ordinary argument for ceding judgment and influence to elite representatives of the *producers* of science and technology—while excluding everyone else who will be affected—is that lay citizens have neither the competence nor the passion to be involved.

Against this argument stands the brute fact that given a chance, our Boston-area Citizens' Panelists, like their counterparts in consensus conferences throughout Europe, competently assimilated a broad array of written and oral expert and stakeholder testimony, and then integrated this information with their own, very diverse life experiences to reach a well-reasoned collective judgment. Their conclusions pass a "reality test"—a groundedness in the daily experience and concerns of everyday people—that expert conclusions routinely fail. To me, this stands as strong evidence for both the need and the practicality of democratizing U.S. science and technology institutions and decisions across the board.

At least in the abstract, we Americans are fiercely proud of our democratic heritage and our technological prowess. But it is striking that we do virtually nothing to ensure that these twin sources of national pride are in harmony with one another. Consensus conferences are not a magic bullet for all that ails democracy or for ensuring that science and technology are responsive to social concerns. But they do reawaken hopes that, even in a complex technological age, democratic principles and procedures can prevail and, indeed, extend into the technological domain.

Acknowledgments

This essay includes material previously published in Sclove (1996) and (1997); it also draws on interviews and correspondence conducted throughout 1994–1996 with Johs Grundahl, Danish Board of Technology, Copenhagen; Anneke Hamstra, Institute for Consumer Research, The Hague; Simon Joss, Science Museum, London (now with the Centre for the Study of Democracy, London); Lars Klüver, Danish Board of Technology, Copenhagen; Lydia Sterrenberg, Rathenau Institute, The Hague; and Norman Vig, Carleton College, Northfield, Minnesota. To receive future updates on the author's activities and publications involving the democratization of science and technology, including information about upcoming U.S. consensus conferences, subscribe to receive the free online newsletter *Loka Alerts* by sending an e-mail request to <Loka@loka.org>. Previous Loka Alerts are archived on the Loka Institute web pages at <www.loka.org>.

Notes

1. For other recent participatory innovations in science and technology decision-making see, for example, Sclove (1992); Renn et al. (1995); and Sclove, Scammell, and Holland (1998).

2. A growing list of the consensus conference that have been organized by the Danish Board of Technology may be found on the World Wide Web at <http://www.tekno.dk/eng/>.

3. I have discussed the importance of democratic community and its relationship to technology in Sclove (1995); see also Putnam (1996).

4. For more information about the organization of European consensus conferences, see Joss and Durant (1995).

5. OTA (1988, p. 85). This and other OTA publications are available via the World Wide Web at <http://www.wws.princeton.edu/~ota/>.

6. Lydia Sterrenberg, The Rathenau Institute, The Hague, The Netherlands and Anneke M. Hamstra, Institute for Consumer Research, The Hague, The Netherlands interviewed together by Richard Sclove, October 11, 1994. The lay panel report of the Dutch consensus conference on genetically modified animals is Public Debate (1993).

7. Simon Joss, telephone interview with Richard Sclove, July 14, 1995; see also Joss (2000).

8. Lars Klüver, Director, Danish Board of Technology, personal communication August 2, 1995.

9. We called the process a "Citizens' Panel" rather than a "consensus conference" for two reasons. First, the U.S. National Institutes of Health (NIH) already use the term "consensus conference" to refer to a process that seeks

consensus among medical experts, rather than among laypeople (e.g., Veatch, 1991); thus we wanted to avoid confusion with the expert-based NIH process. Second, we have found that the term *consensus* can be unnecessarily controversial in the U.S. political context, where for some people it harbors connotations of group or state suppression of individual points of view.

10. The nonprofit Jefferson Center in Minneapolis (on the World Wide Web at <www.usinternet.com/users/jcenter>) has organized related "Citizen Jury®" processes on several complex social issues, including national health policy and priority-setting in the federal budget.

11. One year after the Boston consensus conference, a first Swiss consensus conference was organized on the topic of "Electricity and Society" (PubliForum, 1998). While the Swiss lay panel was socioeconomically and racially less diverse than the Boston lay panel, it broke important new procedural territory in selecting lay citizens who spoke three different languages (French, German and Italian), thus necessitating the use of simultaneous interpreters and translators. This Swiss event thus opens intriguing possibilities for the future practicability of organizing consensus conferences transnationally; see also Andersen 1995.

12. Project manager Laura Reed and her assistant, Kerri Sherlock, were terrific, as were our professional facilitators, Kagan Associates of West Newton, MA.

13. See the final report "UK National Consensus Conference" (1994). For critiques of the first British consensus conference, see Purdue (1996) and Joss (2000).

14. See also Guston (1999) for an independent evaluation of the Boston-area Citizens' Panel.

15. Despite this shortcoming, however, my guess is that the lay panelists knew more than the average U.S. congressman who voted on the Telecommunications Reform Act in 1996. Members of Congress must focus on so many issues at a time that few are able to provide the attention to any particular issue that the panelists in Boston gave.

Bibliography

Andersen, I.-E. (Ed.). (1995). *Feasibility study on new awareness initiatives: Studying the possibilities to implement consensus conferences and scenario workshops.* Luxembourg: Directorate-General XIII/D-2, European Commission.

Bimber, B., & Guston, D. H. (Eds.). (1997, February–March). Technology assessment: The end of OTA. Special Issue of *Technological Forecasting and Social Change, 54*(2–3), 125–286.

Consensus Conference. (1989). Consensus conference on the application of knowledge gained from mapping the human genome: Final document. Copenhagen: Danish Board of Technology.

Consensus Conference. (1992). Consensus conference on technological animals: Final document (preliminary issue). Copenhagen: Danish Board of Technology. Also available on-line: <http://www.tekno.dk/eng/publicat/92teaneo.htm>.

Consensus Statement. (1997). Consensus statement of the citizens' panel on telecommunications and the future of democracy, April 4, 1997. [on-line]. Available: <http://www.loka.org/pages/results.htm>.

Cronberg, T. (n.d.). Technology assessment in the Danish socio-political context. Technology Assessment Texts No. 9. Lyngby, Denmark: Unit of Technology Assessment, Technical University of Denmark.

Dickson, D. (1988). *The new politics of science.* Chicago: University of Chicago Press. (Reprinted with new preface; original work published 1984)

Flint, A. (1997, April 5). At Tufts, No wonks need apply: Citizens panel formulates policy. *The Boston Globe*, September 3–6, pp. B1, B6.

Guston, D. (1999, Autumn). Evaluating the first U.S. consensus conference: The impact of the citizens' panel on telecommunications and the future of democracy. *Science, Technology and Human Values, 24,* (4), 451–482. Also available on-line: <http://policy.rutgers.edu/papers/>.

Hackman, S. (1997, August/September). First line: And now a word from your neighbors. *Technology Review, 100,* (6), 5.

INRA. INRA (Europe) and European Coordination Office SA/NV. (1991, June). *Eurobarometer 35.1: Biotechnology.* Brussels: European Commission; Directorate-General; Science, Research, Development; "CUBE"— Biotechnology Unit.

Joss, S. (2000). Participation in parliamentary technology assessment: From theory to practice. In N. J. Vig & H. Paschen (Eds.), *Parliaments and technology: The development of technology assessment in Europe* (pp. 325–362). New York: State University of New York Press.

Joss, S., & Durant, J. (Eds.). (1995). *Public participation in science: The role of consensus conferences in Europe.* London: Science Museum.

Klüver, L. (1995). Consensus conferences at the Danish Board of technology. In S. Joss & J. Durant (Eds.), *Public participation in science: The role of consensus conferences in Europe* (pp. 41–49). London: Science Museum.

National Science Board. (1998). *Science and engineering indicators—1998.* NSB-98-1. Arlington, VA: National Science Foundation. Also available on-line: <www.nsf.gov/sbe/srs/seind98/start.htm>.

OTA. U.S. Congress, Office of Technology Assessment. (1988, April). *Mapping our genes— Genome projects: How big, How fast?* OTA-BA-373. Washington, DC: U.S. Government Printing Office.

Public Debate. Public debate: Genetic modification of animals, Should it be allowed? (1993). The Hague: Netherlands Office of Technology Assessment.

PubliForum. PubliForum "Electricity and Society," (1998, 15–18 May). Bern: Citizen Panel Report. Bern, Switzerland: Technology Assessment Programme, Swiss Science Council.

Putnam, R. D. (1996). The strange disappearance of civic America. *The American Prospect, 24,* (Winter), 34–50.

Renn, O., Webler, T., & Weidemann, P. (Eds.). (1995). *Fairness and competence in citizen participation: Evaluating models for environmental discourse.* Dordrecht, Boston and London: Kluwer Academic.

Sclove, R. E. (1995). *Democracy and technology.* New York and London: Guilford Press.

Sclove, R E.. (1996, July). Town meetings on technology. *Technology Review, 99,* (5), 24–31.

Sclove, R. E. (1997). Citizen policy wonks. *Yes!: A Journal of Positive Futures, 3,* Fall: 52–54.

Sclove, R. E. (1998, February 27). Better approaches to science policy, *Science, 279,* 1283. Also available on-line under the Publications section of the World Wide Web pages of the Loka Institute at <www.loka.org>.

Sclove, R. E., Scammell, M. L., & Holland, B. (1998, July). *Community-based research in the United States: An introductory reconnaissance, including twelve organizational case studies and comparison with the Dutch science shops and the mainstream American research system.* Amherst, MA: The Loka Institute. Also available on-line as a free download via the World Wide Web pages of the Loka Institute at <www.loka.org>.

Technology and democracy: The use and impact of technology assessment in Europe. (1992, November 4–7). *Proceedings of the 3rd European Congress on Technology Assessment, Copenhagen.* 2 vols. Copenhagen: Teknologi-Naevnet (Danish Board of Technology).

UK national consensus conference on plant biotechnology: Final report. (1994). London: Science Museum.

Veatch, R. M. (1991). Consensus of expertise: The role of consensus of experts in formulating public policy and estimating facts. *Journal of Medicine and Philosophy, 16,* 427–445.

Walker, R. S. (1995, March 2). Democratizing R&D policymaking. Lecture and discussion presented at the 10th Annual Meeting of the National Association for Science, Technology and Society, Arlington, VA.

3

Democratizing Agricultural Knowledge Through Sustainable Farming Networks

NEVA HASSANEIN

Introduction

In 1941 Carl C. Taylor, a prominent rural sociologist of the day, wrote a pamphlet for the U.S. Farm Security Administration in which he encouraged farmers to visit with their neighbors and systematically "trade ideas." Taylor reasoned that neighboring farmers could collectively improve on their "common sense" about farming practices by sharing their experiences and thoughts with one another (1941, p. 6). He also suggested that by talking things over and coming to shared understandings of the problems they wrestled with, farmers could generate completely new ideas and strategies for farming. Once "a good idea" is discussed in terms of the "actual experiences of the people who are going to use it," he wrote, "it has been put through the sieve, so to speak. . . . It is thus refined right down to the place where it will work" (1941, p. 6). For Taylor, such exchanges were much more than utilitarian; they were key to the successful functioning of democracy. He concluded his pamphlet by writing:

> We believe in democracy because we believe every individual has a contribution to make to the solution of our common problems. A meeting of neighbors and friends in their homes is the grass roots of democratic organization, and a trading of ideas among neighbors is the way to make democracy work. (Taylor 1941, p. 7).

A half a century later, the "common problems" in agriculture are so profound that analysts and activists have come to question how long the system can be sustained economically, socially, and environmentally. Many

49

have argued that the roots of unsustainability can be traced, in part, to the epistemological underpinnings, institutional structure, and resulting technical trajectory of agricultural science. Not surprisingly, then, the formal institutions of agricultural science have had little to offer those who seek to adopt alternative practices and ideas that deviate from the conventional agricultural paradigm (Kloppenburg, 1991). As a result, farmers and rural advocates within the contemporary sustainable agriculture movement have turned to one another for the new knowledge they need and thus have taken up Taylor, albeit unwittingly, on his suggestion of trading ideas.

In many ways the grassroots foundation of the contemporary sustainable agriculture movement can be found in the farmer-to-farmer networks that have developed in many areas of the country (Bird, Bultena,& Gardner, 1995). Within these loosely structured organizations, farmers typically share their own experiential or "local" knowledge about sustainable farming, unearthing their own capacities to generate and share new ideas. The ramifications are perhaps much greater than Carl Taylor might have imagined. In these democratic organizations, sustainable agriculturalists fundamentally challenge the inequitable power relations characteristic of the dominant system of agricultural knowledge production and distribution.

This chapter describes and analyzes several key elements of the democratization of agricultural knowledge in the sustainability movement. It begins with a review of the now familiar charge that, as currently constituted, the formal institutions of agricultural research and extension are not democratic. I go on to illustrate some of the prominent dimensions of the alternative knowledge system that has developed largely outside of agricultural scientific institutions and at the local level. In particular, I draw on qualitative research I conducted in two sustainable farming networks in Wisconsin over a two-year period in the early 1990s (Hassanein, 1999). One is a network of dairy farmers practicing the technique of intensive rotational grazing, and the second is a network of women in sustainable agriculture. The two networks were organized for quite different purposes, but each illustrates how democratic participation in the production and exchange of agricultural knowledge is fundamental to a truly sustainable farming system.

From Critique to Innovation

Over the course of the last century, the food and agriculture system has become increasingly industrialized and dominated by multinational agribusiness corporations (Bonanno, Busch, Friedland, Gouveia, Mingione, 1994). If there is a common conviction within the sustainable agriculture movement, it is opposition to these trends (Bird, Bultena, & Gardner, 1995). Skeptical about the long-term ability of the existing system to sustain itself and those who depend on it, activists have sought to demonstrate that the bountiful harvest of industrialized agriculture has adverse ecological, social, and economic consequences that are destructive to both people and the land. While this "unsustainability" of conventional agriculture is generally agreed on within this social movement, precisely what constitutes "sustainability" is a subject of ongoing discussion. A useful and inclusive definition seems to be an agricultural system "that equitably balances concerns of environmental soundness, economic viability, and social justice among all sectors of society" (Allen, Van Dusen, Lundy, & Gliessman, 1991, p. 37).

In an attempt to understand the origins of what has made agriculture unsustainable in the first place, activists have critically examined the role of the formal institutions of agricultural research and extension in promoting the industrialization of agriculture. In particular, sustainable agriculture advocates and sympathizers from within the research community have focused much of their attention on the land-grant system as opposed to corporate research (e.g., Berry, 1984; Kloppenburg, 1988). The foundations for this publicly funded system were established through a series of federal and state legislative actions, beginning in 1862 and continuing during the following 50 years. The triad system of agricultural science includes conducting research at agricultural experiment stations, teaching scientific principles of agriculture, and disseminating scientific knowledge through the county extension agent system to people in the countryside. Because the land-grant system was created and is supported largely with public funds, advocates of sustainable agriculture have argued that it *should* therefore be relatively more accountable and receptive to public input than corporate research programs (Thornley, 1990).

Critics of agricultural science in the United States have argued that this system has not been democratic in that opportunities for broad, public participation have been lacking in at least two key areas. First, the superiority of scientific over farmer-generated knowledge has been claimed and quite widely accepted. In contrast, for thousands of years it had been farmers and craftspeople who generated the knowledge that shaped the practice of farming. That changed when the land-grant universities were created and expanded. In alliance with "gentleman farmers" who valued scientific training, core pressure groups of scientists, journalists, and industrialists argued that increasing agricultural production through science was essential to moving the national economy toward industrialization (Kirkendall, 1987). In this process of establishing legitimacy and securing public funding for agricultural science, farmer-generated knowledge was maligned and slowly hidden from history (Danbom, 1986). As John Bennett (1986) demonstrated, the perception that scientists know more than farmers established a unidirectional flow of communication from expert to practitioner. With farmers viewed as the recipients rather than the generators of knowledge, jurisdictional boundaries were set around whose knowledge would be considered legitimate (Marcus, 1985). This undemocratic tendency undermines sustainable agriculture because sometimes farmers' own experiential knowledge may have more validity and immediate utility than scientific knowledge, as we shall see below (Gerber, 1992; Kloppenburg, 1991).

A second way in which broad, public participation has been lacking in agricultural science is that the research questions that have been asked have tended to represent the interests of certain members of society over others. Specifically, agribusiness corporations and large-scale, industrial farmers have had a disproportionate influence over the research topics that have (and have not) been pursued, and these interests have been the primary beneficiaries of the technologies generated as a result (see Buttel, Larson, & Gillespie, 1990 for a review). Agribusinesses have profited by supplying the seeds and technologies (e.g., pesticides, machinery) that are required by the conventional system. As more and more capital has become necessary to farm, it has been harder and harder for small and medium-size farms to keep pace with the technological treadmill, and many have thus been forced out of farming altogether (Berry, 1977; Strange, 1988).

In addition, women and people of color have had little influence over the directions that agricultural research has taken, as evidenced by the astounding lack of diversity within agricultural research communities (Sachs, 1983; Schor, 1996). As William Lacy (1993) observed, there is a general tendency to neglect those who are separate from people in research institutions, whether by physical space, socioeconomic status, time (i.e., future generations), gender, or ethnicity. Lacy (1993, p. 42) concluded that "the broad experiences and innovative ideas required to make research programs responsive to a sustainable agriculture agenda may be lacking."

The observations made above contrast with the classic view of science that scientists should operate in institutions free from external manipulation and independent of social and political pressures. Such neutrality has been the basis for claiming the objectivity and thus the superiority of scientific knowledge. However, neutrality has clearly not been achieved, and scientists have too often failed to ask questions in the broader public interest.

Not surprisingly, then, sustainable agriculturalists have found that agricultural science has generally had little to offer them as they pursue alternative farming techniques and marketing systems. One important strategic response to this state of affairs has been to try to reorient agricultural science toward lines of inquiry and research methods that are more useful to sustainable agriculture (see Buttel, 1993). These efforts have successfully increased public funding for an alternative agriculture research agenda and created new institutional arrangements at a number of land-grant universities (Gardner,

1990). For example, advocates persuaded Congress to establish the Sustainable Agriculture Research and Education program in 1985, which has since provided financial support for hundreds of projects aimed at reducing the ecological impacts of conventional farming techniques and increasing farm profits. In addition, some farmers, activists, and academics have advocated for and experimented with new models for conducting and participating in agricultural research (see Lockeretz & Anderson, 1993). These new models tend to emphasize multidisciplinary research on agroecological questions, and often involve farmers in the research process either through on-farm experimental trials or in an advisory capacity. Thus, activists have recognized that democratizing agricultural science not only requires redirecting public research dollars toward alternatives, but also changing how that research is undertaken.

Academics and activists attempting to reform agricultural science policy represent just one arm of the sustainable agricultural movement. An equally important response has involved the development of an alternative knowledge system that operates largely outside of the institutions of agricultural research and extension. Specifically, farmers and activists have created a host of rural organizations that emphasize farmer-generated knowledge, promote holistic and ecological thinking, and embrace practical research (Bird, Bultena, & Gardner, 1995). In some instances, this alternative knowledge system involves scientists working in research institutes that seek to reflect a vision of a more socially responsbile science, such as the Land Institute in Kansas (Jackson 1990). Perhaps, however, greater potential for realization of a democratic approach to knowledge generation and dissemination lies with local sustainable agriculture organizations that are primarily farmer-based. Some of these groups conduct their own on-farm experiments into alternative farming methods with assistance from maverick scientists at land grant universities. Practical Farmers of Iowa is a good example of such a group that has tried to bridge the gap between farmers' unreplicated field observations on the one hand, and small, unrepresentaive plots conducted at university field stations on the other (Rossmann, 1994).

By contrast, other groups are networks or clubs where farmers share knowledge they have generated through experience and personal observation rather than through systematic experimentation. This latter strategy for democratizing knowledge is reflected in the two case studies of organizations represented below. As mentioned above, one group is made up of dairy farmers who have adopted a technique known as intensive rotational grazing, and the second group is for women in sustainable agriculture. The analysis is based on participant observation I conducted from late 1992 until the spring of 1995, as well as on 22 interviews with 11 of the most active members of each network. As I will discuss below, farmers in these networks draw on their own knowledge and extend that knowledge by sharing ideas, innovations, and techniques among a wider community. By "knowledge" I refer both to substantive or technical information about specific topics and to the assumptions that shape how such information is constructed and exchanged. This alternative knowledge system suggests a democratic model that (1) draws on the problem-solving, creative capacities of farmers; and (2) involves the horizontal exchange of ideas and techniques that can undergird progress toward sustainable agriculture.

Rotational Grazing and Local Technical Knowledge

Although some agricultural scientists in the United States experimented with putting cows out to pasture in commercial dairying in the 1950s, they concluded that production could be maximized by utilizing land for intensive forage crop production and feeding that forage to cows kept in confinement throughout the year. Since that time, dairy scientists have pursued research questions almost exclusively related to such "confinement feeding," a system that requires major capital outlays for farm inputs (e.g., huge silos, harvest machinery, pesticides) and that is associated with a number of environmental problems. Thus, in the late 1980s when farmers in Wisconsin began to experiment with intensive rotational grazing as a technological alternative, they found that agricultural researchers did not have much useful information to offer them. An agronomist at the University of Wisconsin stated the problem succinctly: "I get asked more questions that I can't answer than I can answer about rotational grazing . . . There's been no research done."

The new "grass farmers" turned to one another for the knowledge they needed and formed grazing networks to facilitate that knowledge exchange process. The Ocooch Grazers Network, which I studied from 1992–1995, was one of many networks in Wisconsin organized by and for dairy farmers practicing the technique that is considered heresy in the United States but that has been the norm in a number of other countries for decades. The main organizational activity of these networks is pasture walks that are held at a different member's farm each month during the grass season. During the period of my study, most of the 100 or so members of the network lived in two neighboring counties in southwestern Wisconsin. Fifteen men and five women were among the core membership who routinely attended events while I was a participant observer.

Unlike confinement feeding where the operator feeds cattle confined to the barnyard and surrounding area, rotational grazing involves intensively managing the land as permanent pasture so that livestock can harvest high-quality forage for themselves for as much of the year as possible. Grazing involves dividing land into small areas and rotating animals through these "paddocks" according to the time necessary for proper regrowth and recovery of the plants. In describing the shift to rotational grazing, one network member said plainly: "For years, I fought and borrowed to bring feed to the cow. Then I finally figured out that I could bring the cow to the grass."

Although this technological reversal sounds simple, questioning conventional wisdom and actually learning how to effectively "bring the cow to the grass" proved to be quite a difficult task for members of the Ocooch Grazers Network. What is critical for successful grazing is learning new ways of seeing and thinking, rather than buying or understanding how to use new tools. Spending time with grazers and attending Ocooch pasture walks, I frequently heard farmers stress the importance of learning from their own "experience" and "observation," rather than from "recipes." For example, in response to one farmer's question about how much pasture land to give dairy cows each day, another network member replied: "You have to look ahead and determine when to graze and that requires experience and observation. There aren't any rules or regulations. You gotta look at the pasture. You can't look at a calendar and know what to do. The whole thing is observation."

The network members reported that the practice of grazing very much depends on the development of local knowledge. Local knowledge is the practical skill that develops with personal experience of and careful attention to the unique, yet changing, social and physical features of a particular place or activity (Kloppenburg, 1991). The grass farmers' emphasis on grazing as a "thinking process" applied to a specific place contrasts with Wendell Berry's (1984) characterization of conventional agricultural practices as a set of prescriptions that can be applied anywhere without a detailed understanding of the local agroecosystem. In other words, grazers asserted the validity of their own local knowledge and the vital role it plays in the success of this sustainable agricultural practice. In this way, members of the Ocooch Grazers Network democratized the *production* of agricultural knowledge. Knowledge created by a few research scientists and handed down to farmers was not appropriate, in large part because of the importance of continually evaluating a range of variables on a specific farm at a particular time. Farmers themselves were the best source of the specific information and ideas they needed. Thus, while the grazing network developed because of the paucity of information on how to practice rotational grazing in the Upper Midwest, what resulted was a fundamental assertion of the validity and utility of farmer-generated knowledge.

Previous analyses of local knowledge in sustainable agriculture have tended to emphasize the deeply personal or tacit character of this way of knowing (Kloppenburg, 1991). However, social movement theorists have recently recognized that such personal knowledge is transmissible and can be socialized. For example, in an analysis of several "new social movements,"

Hilary Wainwright (1994) argues that such movements are characterized by a democratic approach to knowledge that involves the sharing and combining of individuals' personal knowledge to further collective goals.

This democratization of the *dissemination* of knowledge was evident in the Ocooch Grazers Network. Although the network members recognized that they must each individually develop their own local knowledge of rotational grazing, these grazers valued learning from others who had direct experience using the technique. With the following statement, one network member captured many of the central features of knowledge exchange within the Ocooch Grazers Network:

> The main purpose is to obtain information and share information. There is a pasture walk once a month during the grazing season. We go to the host farm to see what's growing there, what's not growing there, whatever they want to show us. . . . I like to go see what ideas people have come up with. There are always new creative ideas. You can see what others are doing to cut costs. Farmers are experts at coming up with cheap, innovative ways to do things. And there are always new circumstances, not necessarily problems, but different things you have to handle. It is useful to talk to people about how to handle the wet spell or how to handle the dry spell. . . . There are always new questions. Chances are someone else has answers [and you] learn from someone who has had firsthand experience.

Indeed, active members of the network constantly shared with each other their technical or substantive knowledge about how to graze, covering such topics as how to develop and improve the pasture, and how to meet the nutritional and energy needs of lactating dairy cows. Often the ideas shared reflected what was learned from trial and error. And as one farmer put it: "I don't want to make mistakes, so why not learn from people that have. So another key thing there [in the network] is learning not only from your mistakes, but other people's mistakes." Network events not only provided an opportunity to hear what had or had not worked for others, but also to see firsthand what others were doing on their farms.

The knowledge exchange process had an egalitarian character to it. Even though the skills of individual network members obviously varied, there was a reluctance among the individuals with greater knowledge to take on the role of "expert." One of the network leaders, for example, had been grazing longer than most other members in the network and had demonstrated considerable grazing skill. He made a clear distinction between making specific recommendations and sharing experiences:

It is satisfying to me to know that I can help someone understand [intensive rotational grazing] a little bit better. . . . But I do not want to get in a situation where I think I know everything, where I'm going to tell you how you have to do this. That makes me an extension man, see, if I'm going to come out and tell you how to farm. . . . It's a situation where a group of equal people come together, and we share stories, and we share experiences, and we share failures, and we share weaknesses. It's a situation where we can share all of these experiences and then use those experiences to build on our own knowledge base and make that knowledge that we've gained conform in some way to what we have in reality out here on our own farms.

By emphasizing sharing and equal participation, this approach rejects the unidirectional flow of information from an expert to a nonexpert. It asserts the validity of experiential knowledge, while simultaneously refusing any claims to an all-knowing expertise.

Participation in the network was not completely "equal," however. Gender seemed to account for some distinct differences in members' interactions. As one woman member said plainly: "Usually men do the talking." It is not that women did not have knowledge to share. For example, they actively exchanged their experiences with techniques for feeding young calves when that topic became a focus of discussion among several women. But in the larger group, women's participation was typically less than equal to that of men. (The second half of this chapter will examine gender inequality in more depth.)

Overall, an important characteristic of the information exchange within this network was that members *themselves* asked what the appropriate questions were. In this way, the important questions were identified democratically with farmers participating and sharing together. Another important characteristic of the network is that the farmers looked to *each other* for answers. In so doing, they often drew on their own personal, local knowledge, which they shared with others who were similarly situated. The evidence suggested that local knowledge was repeatedly extended beyond the individual and meaningfully shared with others interested in intensive rotational grazing so as to increase members' chances of realizing their goals. Once a network member articulated in the network setting an idea or observation derived from their own personal knowledge, that knowledge then became a social product available for use and interpretation by a community of knowers. By recognizing the limits of expertise, the network also suggested the possibility of a knowledge system that values the unique contributions of many—rather than a few—different experiences. This pluralistic and egalitarian approach to the collective creation and horizontal exchange of ideas illustrates how the activities of this network served to democratize agricultural knowledge.

Gender, Social Location, and Knowledge Exchange

Women are certainly not new to agricultural production, but women's various roles in agriculture have changed over time. As recent historical studies have documented, the research and extension activities of land-grant universities helped to institutionalize and solidify gender relations on family-owned and -operated farms, such as those that predominate in Wisconsin today (Jellison, 1993; Neth, 1995). Specifically, over the course of the first half of the 20th century, a division developed between the scientific knowledge about agricultural production that was delivered to male farmers on the one hand, and "domestic science" on "homemaking" that was delivered to farm women on the other. In this way, the institutions of agricultural research clearly differentiated between the "farmer" and the "farmer's wife" in the delivery of information. This separation has not been equal, however, with disproportionately greater staff and funding resources being devoted to information about agricultural production (Knowles, 1985). And, perhaps even more important, the message that farming is a male occupation and that women's work is secondary and supportive was conveyed by agricultural research and extension institutions through the unequal and separate delivery of knowledge to farm families (Sachs, 1983).

It is within this context that the Wisconsin Women's Sustainable Farming Network was formed by and for women in 1992. The impetus for forming the group came from a couple of women who had already been involved in another sustainable agriculture network and who noticed the lack of active participation by other women farmers. As an experiment, they organized a conference workshop aimed solely at women in agriculture, and found, in the words of one organizer, "It was a totally different atmosphere. The women felt a lot of support and wanted it to continue." The organizers also recognized that "there are things we know we can teach each other and learn from each other."

By 1996, the membership list had grown to nearly 300. The network is a statewide group that meets several times a year for one to two day conferences. In defining their organizational mission, the members themselves captured the prominent themes that ran through their conversations and activities:

> Our mission is to inspire women farmers with a strong support network that promotes successful sustainable farming. We will share personal experiences, technical information, and marketing strategies.

This mission contrasted with the rotational grazing network where members were united primarily by a particular sustainable farming technique they all used. Rather, members of the Women's Network pursued a range of farm enterprises and engaged in a variety of farming practices—from sheep dairying to cut flowers to market gardens. Although the women shared an intense interest in sustainable farming, members of this network were united primarily by a common social location, their gender identity.

Members of the Women's Network democratized knowledge about the social relations in agriculture by developing a shared understanding of each member's personal knowledge of gender discrimination in farming. If identifying as a grass farmer constituted a technological reversal from the dominant methods of dairy production, a woman identifying herself as a sustainable farmer constituted a kind of social reversal from the dominant understanding of who can be a farmer. No less than the grazer who needed knowledge to make that technological reversal possible, so too the members of the Women's Sustainable Farming Network needed knowledge to make that social reversal possible.

Like all knowledge, the production of local, personal knowledge for sustainable agriculture is generated from a particular and partial perspective (Harding, 1991, Feldman & Welsh, 1995). Therefore the content of the knowledge exchanged among network members was shaped both by the diversity within the group and by their shared gender identity. The members lived experiences differed from one another in terms of age, living arrangements, weather, farm type and location, and length of time spent in agriculture. This diversity provided a rich source for the knowledge shared within the group. Despite this diversity, however, there was a common thread woven through much of the knowledge shared in network events: overcoming the ways in which the knowledge and labor of farm women has been devalued in contemporary agriculture and recognizing these women's identities as farmers in their own right. As sociologist Dorothy Smith (1987, p. 78) argued, women may differ considerably in their lived experiences, but what women "have in common is the organization of social relations that has accomplished our exclusion."

From this standpoint, members of the women's network generated and exchanged their own personal knowledge related to certain institutional impediments and social constraints that are often deeply embedded in rural communities. For example, Women's Network members' encounters with gender stereotyping in their everyday lives often meant that their own farming activities were rendered invisible. One member wondered why "Boots and gloves aren't made for women. Don't they think women work?" Another woman cited an instance when she went to buy a tiller for her market garden, and rather than talk to her, the salesman insisted on addressing her husband who had gone along. She felt that these kinds of experiences can happen in "subtle ways" that can still be "pretty intimidating." Another member said that in many settings, she has not felt that she "belonged there," and "even going down to my local feed co-op, I always feel like they look at me like I'm from Mars. . . . I don't know if it's just our specific area, but most of the time I don't feel really treated with respect, that's just been my experience." One member summed up the experiences of many of the members by saying: "It's really hard being female in a man's occupation or what's traditionally been a man's occupation. . . . It's just the women aren't valued for what they do is what it comes down to."

These kinds of experiences in a gendered society were a principal source of personal knowledge that members drew on in their network exchanges. Like gender discrimination everywhere, gender relations in agriculture can be experienced by a woman without her thinking of it as such. A deliberate process of sharing personal experiences and analyzing the recurrence of those experiences to find common causes within social arrangements is one way that the organization of social relations can be better understood. The "consciousness raising" groups organized by feminists in the 1970s constitute an explicit example of such collective reflection on the political meaning of personal experiences (Ferree & Hess, 1985). Although members of the Women's Network differed in the extent to which they identified as feminist, women in the network shared certain ways of seeing, knowing, and understanding farming experiences that seemed to them to differ from those of their male counterparts. Discussing with one another their personal struggles with gender stereotypes and relations in agriculture led them to understand that they operated within a context that discouraged them from farming.

At the same time, however, these women did not necessarily see social constraints as all-encompassing or all-powerful. Rather, an important element of some network members' personal knowledge seemed to be their own understandings of themselves as individuals who could make choices and take actions on their own behalf. Moreover, in learning about the skills and achievements of other women who have struggled with and often overcome obstacles related to existing gender relations in agriculture, members seemed to obtain a crucial piece of knowledge: They too might be able to succeed as farmers. Indeed, the members themselves identified the fact that they found "role models" in the network to be a particularly valuable function of the group.

One principal way in which this role model function seemed to manifest itself within the network was the considerable amount of time members devoted to the "introductions" that customarily began each meeting or conference. The organizational practice began with the day-long conference in the spring of 1993. The leaders invited the sixteen participants to "share your stories . . . where you are from, history in farming, what you are doing now, your dream for the future, any words of wisdom or helpful hints you've learned along the way, and what you would want from this group." These "introductions" took two hours as women addressed some of these questions and were often peppered with further questions and comments from others. Subsequent meeting agendas always included sufficient time for this sharing because of the strong appreciation that members expressed for hearing each other's stories. Such sharing meant that members seemed to teach one another by example. That is, they articulated their commitments to their pursuits and displayed their competence for others to see. In this way, these women farmers surreptitiously exchanged the knowledge that, as one member put it succinctly, "you can do it."

In addition to democratizing knowledge about *who* can be a farmer, a substantial portion of network meetings were devoted to the exchange of specific ideas about *how* to overcome gender-related obstacles and be success-ful at farming. In these exchanges, the important research questions were identified democratically by network members themselves, not by a few experts at agricultural institutions. And network members looked to each other for answers. For instance, one member advised others on how to obtain a loan from a bank that only wants to lend to one's husband. Another member explained how she coaxed her cows into doing what she wants rather than physically forcing them as men, whose bodies are generally bigger and stronger than hers, tend to do.

This democratic exchange was not limited, of course, to gender-related issues, because these women did not separate their practical farming inter-ests from their identities as women. Accordingly, members of the network also shared their knowledge of specific techniques related to sustainable farming, including, for example, how to set up fencing systems for rotational grazing or how to run a successful market garden. Like the grazers, members of the Women's Network stressed the value of learning from the practical experience of others. As one member said:

> You can get the theory from the books. But it's seeing it at the field days or hearing it and being able to ask questions about it that really helps you to understand it yourself and be able to bring it back to your farm and do it.

Even while network members were engaged in very different kinds of enter-prises, they appreciated the opportunity to hear what other women were doing on their farms. It exposed them to new ideas that they may not have previously considered incorporating into their own operations. This diversity also served to create a more egalitarian character to the knowledge exchange process, as this member explained: "The thing that diversity does is it doesn't create any kind of hierarchy. There's no best grazer, there's no best flower grower, because we're all so diverse. . . . We're kind of equal. . . . Each person who's coming in there is the expert in what they are doing."

In some cases, the process of democratizing knowledge consisted not of creating and exchanging ideas unfamiliar to those in agricultural research institutions, but rather of disseminating knowledge that has been systematically denied to women because of gender discrimination regarding acceptable interests. For example, a number of women recognized that they had been discouraged from mechanical interests and thus had little mechanical knowledge. To help address their identified needs in this regard, steering committee members organized conference workshops on such topics as "selecting and maintaining tools and machinery" and "farm carpentry." In this case, knowledge readily available and familiar to men but not women—thus making men privileged with respect to becoming farmers—was democratized by putting women on an equal footing with respect to such knowledge.

This kind of exchange of technical information was an essential function of the Women's Network; otherwise, these women might have gone to other organizations for rural women. But network members consciously rejected those other groups precisely because the technical aspects were ignored there. At the first meeting she attended, one member made this clear when she introduced herself to the group. Initially, she had been reluctant to come to the meeting, because her previous experiences with groups for rural women had been negative. To illustrate, she relayed a story of the time she had agreed to attend a meeting sponsored by the Homemakers Club, because her neighbor had asked her to go. She explained that the event was a "bad experience," because she learned how to "tie scarves" when she would have rather learned about how to "fix carburetors." The group roared with laughter. Clearly, this woman wanted information about the technical side of farming. But, according to historian Mary Neth (1995, p. 138), the Homemaker's Clubs were organized initially by the extension service in 1914, and ultimately served to alter "definitions of what was appropriate work for farm women" by concentrating on women's work in the home rather than on agricultural production. As a result, the kind of information that would have satisfied this farmer's interests was not incorporated into the Homemaker's Club agenda.

If "knowledge is power," then systematic exclusion of a group of people from knowledge is fundamentally undemocratic. The land-grant universities have helped to exclude women from the knowledge they need to be farmers by institutionalizing gender divisions of labor on family farms. These network members sensed the loss of power resulting from knowledge deficiencies, and a principal thrust of their activities was to overcome those deficiencies. As we have seen, they did that by sharing their experiences and knowledge of gender relations in agriculture that have tended to exclude them, by serving as "role models" for one another and thereby sharing the knowledge that it is possible to overcome that historical exclusion, and by providing the necessary technical knowledge to succeed at sustainable farming.

Conclusion

The farming networks discussed here are but two examples of the organizational mechanisms that farmers and activists have established throughout the country in an effort to generate and disseminate the knowledge necessary for building a more sustainable agriculture. Movement actors have developed this alternative knowledge system as one strategy for responding to the ways that agricultural science and the structure of the land-grant system have contributed to the industrialization, globalization, and corporate domination of agriculture. Underlying these movement activities is a fundamental recognition that knowledge production and dissemination must be more democratic in order to achieve a sustainable agriculture.

Science in the land-grant universities has been characterized by a hierarchical, one-way-street that carries knowledge produced by the few to the many. In each of the networks studied here, however, we have seen examples of farmers' rejection of the assumption that knowledge generation is the prerogative of an isolated, scientific elite. Instead, network members unearthed their own knowledge-generating capacities. They asserted the validity of their personal, local knowledge and the utility of that knowledge in unique physical and social locations.

Agricultural science has also tended to represent certain interests over others. As members of the networks adopted farming techniques consistent with the principles of sustainability, they simultaneously rejected the technical trajectory that has been pursued by scientists and that has benefited agribusiness corporations to the detriment of independent agricultural producers and the land as well. As one member of the Ocooch Grazers Network bluntly put it, "We don't owe agribusiness a living." Members of the Women's Network also rejected the social constraints that have been institutionalized in the land-grant system which has directed knowledge about agricultural production to men while directing knowledge about domestic affairs to women. In each case, network members developed a sense of epistemic self-reliance, as they asked the questions that have not been of interest to agricultural scientists and as they turned to one another for answers to those questions. Moreover, these farmers influenced how the results of their inquiry might be used, based on a set of shared values and a common interest in the sustainability of agriculture.

The activities of the networks suggest the need for greater accountability on the part of agricultural science. This will only be achieved when people can participate more fully in decisions about new knowledge, for themselves and for society. The democratization of knowledge involves equal access to information, as well as equal participation in answering the questions about what knowledge is produced, by whom, for whom, and toward what ends.

Bibliography

Allen, P., Van Dusen, D., Lundy, J., & Gliessman, S. (1991). Integrating social, environmental, and economic issues in sustainable agriculture. *American Journal of Alternative Agriculture 6*(1), 34–39.

Bennett, J. W. (1986). Research on farmer behavior and social organization, chapter 16. In K. A. Dahlberg (Ed.), *New directions in agriculture and agricultural research: Neglected dimensions and emerging alternatives.* Totowa, NJ: Roman & Allenheld.

Berry, W. (1977). *The unsettling of America: Culture and agriculture.* New York: Avon Books.

Berry, W. (1984). Whose head is the farmer using? Whose head is using the farmer?, chapter 2. In W. Jackson, W. Berry, & B. Colman (Eds.), *Meeting the expectations of the land: Essays in sustainable agriculture and stewardship.* San Francisco: North Point Press.

Bird, E. A., Bultena, G. L., & Gardner, J. C. (Eds.). (1995). *Planting the future: Developing an agriculture that sustains land and community.* Ames: Iowa State University Press.

Bonanno, A., Busch, L., Friedland, W., Gouveia, L., & Mingione, E. (Eds.). (1994). *From Columbus to ConAgra: The globalization of agriculture and food.* Lawrence: University Press of Kansas.

Buttel, F. H. (1993). The production of agricultural sustainability: Observations from the sociology of science and technology, chapter 1. In P. Allen (Ed.), *Food for the future: Conditions and contradictions of sustainability.* New York: Wiley & Sons.

Buttel, F. H., Larson, O. F., & Gillespie, G. W., Jr. (1990). *The sociology of agriculture.* Westport, CT: Greenwood Press.

Danbom, D. B. (1986). Publicly sponsored agricultural research in the United States from an historical perspective, chapter 6. In K. A. Dahlberg (Ed.), *New directions in agriculture and agricultural research: Neglected dimensions and emerging alternatives.* Totowa, N.J.: Roman & Allenheld.

Feldman, S., & Welsh, R. (1995). Feminist knowledge claims, local knowledge, and gender divisions of agricultural labor: Constructing a successor science. *Rural Sociology, 60*(1), 23–43.

Ferree, M. M., & Hess, B. B. (1985). *Controversy and coalition: The new feminist movement.* Boston: Twayne Publishers.

Gardner, J. C. (1990). Responding to farmers' needs: An evolving land grant perspective. *American Journal of Alternative Agriculture, 5*(4), 170–173.

Gerber, J. M. (1992). Farmer participation in research: A model for adaptive research and education. *American Journal of Alternative Agriculture, 7*(3), 118–121.

Harding, S. (1991). *Whose science? Whose knoweldge? Thinking from women's lives.* Ithaca, NY: Cornell University Press.

Hassanein, N. (1999). *Changing the way America farms: Knowledge and community in the sustainable agriculture movement.* Lincoln: University of Nebraska Press.

Jackson, W. (1990). Agriculture with nature as analogy. C. A. Francis, C. B. Flora, & L. D. King (Eds.), chapter 14 in *Sustainable agriculture in temperate zones.* New York: Wiley.

Jellison, K. (1993). *Entitled to power: Farm women and technology, 1913–1963.* Chapel Hill: University of North Carolina Press.

Kirkendall, R. S. (1987). Up to now: A history of American agriculture from Jefferson to revolution to crisis. *Agriculture and Human Values, 4*(1), 4–26.

Kloppenburg, J. R., Jr. (1988). *First the seed: The political economy of plant biotechnology, 1492–2000.* New York: Cambridge University Press.

Kloppenburg, J. R., Jr. (1991). Social theory and the de/reconstruction of agricultural science: Local knowledge for an alternative agriculture. *Rural Sociology, 56*(4), 519–548.

Knowles, J. (1985). Science and farm women's work: The agrarian origins of home economic extension. *Agriculture and Human Values, 2*(1), 52–55.

Lacy, W. B. (1993). Can agricultural colleges meet the needs of sustainable agriculture? *American Journal of Alternative Agriculture, 8*(1), 40–45.

Lockeretz, W., & Anderson, M. D. (1993). *Agricultural research alternatives.* Lincoln: University of Nebraska Press.

Marcus, A. I. (1985). *Agricultural science and the quest for legitimacy.* Ames: Iowa State University.

Neth, M. (1995). *Preserving the family farm: Women, community, and the foundations of agribusiness in the Midwest, 1900–1940.* Baltimore: Johns Hopkins University Press.

Rosmann, R. L. (1994). Farmer initiated on-farm research. *American Journal of Alternative Agriculture, 9*(1,2), 34–37.

Sachs, C. E. (1983). *The invisible farmers: Women in agricultural production.* Totowa, NJ: Rowman & Allanheld.

Schor, J. (1996). Black farmers/farms: The search for equity. *Agriculture and Human Values, 13*(3), 48–63.

Smith, D. E. (1987). *The everyday world as problematic: A feminist sociology.* Boston: Northeastern University Press.

Strange, M. (1988). *Family farming: A new economic vision.* Lincoln: University of Nebraska Press.

Taylor, C. C. (1941). Trading ideas with your neighbors. Pamphlet submitted to the U.S. Farm Security Administration. Carl C. Taylor Papers, Collection No. 3230, Rare and Manuscript Collections, Cornell University Library, Ithaca, NY.

Thornley, K. (1990). Involving farmers in agricultural research: A farmer's perspective. *American Journal of Alternative Agriculture, 5*(4), 174–177.

Wainwright, H. (1994). *Arguments for a new left: Answering the free market right.* Oxford, U.K.: Blackwell.

4

Public Participation in Nuclear Facility Decisions

Lessons From Hanford

LOUISE KAPLAN

Environmental movements are often characterized by the organization of citizens around the perception that they are threatened by the production, use, or disposal of toxic materials. Citizens appeal to government officials and scientific experts to remedy these situations on moral or ethical grounds, sometimes relying on the testimony of expert witnesses used for the purpose of providing scientific credibility (Couch & Kroll-Smith, 1997). Citizens come to distrust government officials and scientific experts who fail to support lay claims of the presence of a hazard or who fail to remediate the situation. This occurred in situations such as Love Canal and Woburn, Massachusetts (Levine, 1982; Brown, 1987).

Citizen distrust of government officials and scientific experts may develop because of differing interests among them or because of clashes between the cultural, or lay, rationality used by citizens and the technical rationality used by scientists (Krimsky & Plough, 1988). If the government acknowledges an environmental problem, it may be required to conduct studies, fund and organize the relocation of citizens, or engage in extensive and expensive remediation of a contaminated site. If experts confirm the presence of an environmental hazard, they are subject to review by peers who may dispute scientific methods or conclusions. Lay rationality, the way people evaluate the direct and personal effects of a scientific or technological innovation, is based on trust in the political and democratic processes. Technical rationality, in contrast, appeals to authority and enterprise, trust in the scientific method and rests on scientific norms directed at developing technical solutions to problems. These differing ways of perceiving environmental problems change relationships among citizens, government officials, and experts and lead citizens to seek a role in scientific decision-making.

In this social movement, citizens assert their right to participate in policy decisions that affect their health and safety. Public participation can take many forms ranging from testifying at public hearings to becoming experts about the issues being debated. As participants in public policy, citizens demonstrate the ability to intelligently participate by combining lay rationality with technical rationality. As citizens become experts in their own right, they blend social values and technical data. Both Love Canal and Woburn are well-known examples of citizens conducting studies to support their claims and to affirm their concerns. Furthermore, as citizens gain expert knowledge, they are no longer dependent on government officials or scientific experts and can participate equally rather than being in a less powerful political position.

The Hanford Site, a federal nuclear facility in Washington State, is an example of this social movement model. The federal government established Hanford in 1943 as part of the Manhattan Project to produce plutonium for nuclear bombs. Most of Hanford's operations were cloaked in secrecy that national security policies legitimized. Citizens of Washington patriotically supported Hanford's mission and benefited economically. Hanford stood as a symbol of progress and military might for nearly 40 years.

Beginning in the mid-1970s, various stakeholders questioned the safety of Hanford operations and the possibility of adverse health effects from exposure to radiation. Citizens asserted their right to participate in policy decisions that affected their health and safety and did so intelligently, combining lay rationality with technical rationality. Hanford activists used many different strategies to participate in policy decisions. These included placing citizen initiatives on ballots, becoming experts about radiation and its health effects, testifying at public hearings, legal action, and educating and organizing citizens. As a result of this participation, the U.S. Department of Energy (USDOE) released the Hanford Historical Documents in 1986, which revealed radioactive releases to the environment. Public participation in decision-making about Hanford-related activities continues today.

The emergence of public participation in decisions at Hanford was a process that developed in slow motion over decades. The public's perception of Hanford shifted from a symbol of military might and progress to one of risk. This led to collective action opposing policies that suppressed and withheld information about Hanford. This shift in public perception developed as nuclear technology in general came under scrutiny and citizens questioned the safety of that technology.

This chapter discusses the role of the public in activities that led to revelations about large radioactive releases from Hanford. An historical overview of public participation in the debate about nuclear technology in Washington demonstrates how social problems move through stages. These stages include the public awareness of problems, legitimization of a problem, citizen action, government response, and continued or renewed citizen action. The chapter also describes citizen participation in research studies and public health activities related to Hanford since the release of the Hanford Historical Documents.

Data for this chapter came from scholarly literature, clipping files of libraries, government documents, literature of stakeholder groups, and the personal files of individuals. Data also included 48 interviews of stakeholders who represent people exposed to radioactive releases from Hanford; employees of the United States Department of Energy at Hanford and its contractors, scientists, physicians, and public health officials; local and state government, citizen activists, journalists, and Native Americans.

The First Decades

Hanford developed in secrecy. Thousands of people who built and operated Hanford were unaware of the plant's mission until the United States used atomic bombs in August 1945. Then government officials revealed that Hanford produced plutonium for the first atomic bomb tested in New Mexico as well as for the bomb dropped on Nagasaki, Japan. The secrecy at Hanford that minimized the likelihood of enemies learning about bomb production would come back to haunt government officials decades later.

Secrecy extended to radiation, code named "special hazard" (Weart, 1988). Officials involved in the Manhattan Project were concerned about the potential for adverse health effects from exposure to radiation as well as potential harm to the atmosphere and Columbia River (Smyth, 1945; Hacker, 1987; Sanger & Mull, 1989). The government, however, normalized radiation. An official project report asserted that "the hazards of the home and the family car are far greater for the personnel than any arising from the plants" (Smyth, 1945, p. 149). The government also countered reports of widespread illness and death from the use of atomic bombs in Japan with a public relations campaign about the benefits of the atom (Boyer, 1985). This included a radio program titled "The Sunny Side of the Atom" and a comic book using the character Dagwood Bumstead to convey optimism about the atomic future.

Plutonium production expanded at Hanford following the end of World War II in an effort to stockpile atomic bombs. Hanford remained under a "wartime cloak of secrecy" by order of President Harry Truman (*Seattle Times,*

3/8/46, Bigelow, 1950, p. 33). In the 1950s, national attention turned to the possible hazards of radioactive fallout from atmospheric testing of atomic bombs. Hanford operations, however, remained unaffected. By 1956 Hanford had eight plutonium production reactors, increased from three in 1945, and five chemical processing plants of which there had been two in 1945 (Technical Steering Panel, 1994). Radioactive waste management operations included tanks varying in size from 55,000 to 1,000,000 gallons, which unbeknownst to the public began to leak in the 1950s (Read, 1975). One area farmer commented that during those years:

> I heard it said a thousand times, "What do you think they're doing down there at Hanford?" Well, nobody knows, but was it any concern? "No, They know what they're doing."

In 1962, the United States Congress authorized the use of waste steam to produce electricity from a nuclear reactor to be built at Hanford. President John Kennedy attended the 1963 groundbreaking for the N-Reactor along with 30,000 members of the public, allowed on the nuclear reservation for the first time (Alive, 1983). The public celebration of Hanford's new power plant symbolized the local and statewide support for Hanford. The celebration also reinforced Hanford as a symbol of progress and economic stability. The city of Richland, home to Hanford workers, proudly referred to itself as "The Atomic City."

A year after the N-reactor groundbreaking, a slowdown at Hanford began with three of the eight reactors scheduled for shutdown. Government officials and civic leaders worked to diversify the Hanford area's economy. The State of Washington created the Office of Nuclear Energy Development and authorized the state to acquire land for a nuclear waste disposal site. Washington was "The Nuclear Progress State."

For more than another decade, Hanford had statewide support. Government officials denounced federal threats to cutback operations at Hanford as economically depressing (*Seattle Times*, 1/22/71). In 1972 Washington State's Dr. Dixy Lee Ray received appointment to the Atomic Energy Commission (AEC) and became its chairperson the following year. The Washington Public Power Supply System (WPPSS), a consortium of public and private utilities, began construction of five nuclear power plants, three of which were at Hanford, between 1972 and 1978.

Nationwide, concerns about and opposition to nuclear power developed. Nuclear accidents, questions of legal liability, environmental problems, and high cost eroded support for nuclear power (Lilienthal, 1971; Gamson, 1988; Caufield, 1989). The safety issue blurred the line between nuclear weapons reactors and power plant reactors. If an emergency cooling system failed during a reactor accident, radioactive materials could be released into the air.

Media reports of nuclear problems at Hanford began to appear in the 1970s. Report of a 115,000 gallon leak from a storage tank at Hanford drew a quick response from AEC Chairperson Ray that the leak posed no danger. In 1975 a newspaper article reported at least sixty "unplanned releases" of radioactive materials from Hanford during the prior 30 years. Nonetheless, public support for Hanford remained strong in Washington. Factors that contributed to this support include continued secrecy at Hanford, the federal government's promotion of civilian uses of nuclear technology, and the International Cold War (Boyer, 1985). Finally, Hanford was good for the economy of the Tri-Cities and Washington State.

From Nuclear Progress State to a State Divided

By 1975, less than $100 million of Hanford's $400 million budget was spent on military projects, with the rest being spent on civilian energy research (Connor, 1985). Only one of the nine reactors functioned, along with only one plutonium separations plant and one plutonium finishing plant. Construction of the three WPPSS nuclear power plants was underway.

As the federal government curtailed military operations at Hanford antinuclear power groups grew more active and numerous. In 1976 the first statewide challenge to nuclear technology occurred in Washington, an initiative to safeguard nuclear power reactors, two of which were under construction at Hanford. This initiative was the first of three directed at changing political decisions made regarding nuclear technology. Washington's consensus regarding nuclear technology broke down.

Citizen Initiatives

The Coalition for Safe Energy (CASE) collected over 100,000 petition signatures to place a citizens nuclear safeguard act, Initiative 325, on the ballot in November 1976. The initiative required plant sponsors to provide: information on financial liability in case of accidents, proof of reactor safety systems, a plan for permanent and safe disposal of radioactive waste, and yearly publication of emergency evacuation plans. The initiative's safety theme conveyed the message that nuclear power presented a clear and present danger (Lane, 1976).

Opponents of the initiative constructed a countertheme which was unrelated to the safety issue, "Ban the Ban." The opponents used the slogan to claim the initiative was a moratorium on nuclear power plants since they could not be built under the proposed guidelines. The initiative lost by a vote of 2 to 1. The pronuclear vote was reinforced by the state's election of former Atomic Energy Commission chairperson Dixy Lee Ray as governor.

Supporters and opponents of the initiative agreed, however, that the initiative marked the beginning of attention to nuclear energy in Washington. It also demonstrated the capacity of citizens to research nuclear technology and develop a list of safeguards they believed necessary to ensure safe power plant operations. The initiative did not directly encompass Hanford's main military operations which continued to be politically invisible, cloaked in secrecy. In addition, the temporary storage of Hanford's radioactive waste failed to receive attention even when newspaper articles reported high-level radioactive waste leaks at Hanford's nuclear waste storage site.

High-level nuclear waste created one set of problems at Hanford. A commercial low-level waste site operated by the State of Washington at Hanford created another set of problems. On June 6, 1979, the Yakama Indian Nation tribal council passed a resolution banning the transportation of nuclear waste across the reservation.

Several months later, Washington's Governor Dixy Lee Ray closed the Hanford low-level radioactive waste site after safety problems occurred. The site reopened a month later after new regulations for the truck transport and disposal of waste went into effect (Norman, 1984). A group of activists dissatisfied with Governor Ray's reopening of the waste site at Hanford wanted to ban nuclear waste transportation and storage altogether. In 1980, citizens placed Initiative 383 on the ballot to ban the transport and storage of all but low level radioactive medical waste into Washington until regional compacts were signed. The initiative's slogan, "Don't Waste Washington," reflected its supporters' intent to prevent Hanford from becoming a major radioactive waste disposal site. Initiative supporters saw their efforts as a means of community action and their democratic right to citizen participation in government:

> Because of the potential hazards involved in the transportation and burial of radioactive wastes, it is our belief that the citizens of Washington should have a voice in such an important decision. (Lane, 1980, p. A8)

After the initiative passed by a margin of three to one, a Federal District Court judge ruled it unconstitutional (Whitely, 1981, p. A1). The ruling was eventually upheld by the U.S. Circuit Court of Appeals.

Attention quickly moved from nuclear waste back to nuclear power. Allegations of cost overruns, safety problems, and corruption swirled around the Washington Public Power Supply System (WPPSS) project to build five nuclear power plants in the state. With the WPPSS project (commonly referred to as "Whoops") on the verge of financial disaster, antinuclear activists organized another initiative campaign. Activists placed Initiative 394, the "Don't Bankrupt Washington" campaign, on the ballot in 1981. It passed overwhelmingly and prohibited WPPSS from issuing new public revenue bonds after

July 1982, without a public referendum (Unsoeld, 1983). In 1984, WPPSS plant Number 2, located at Hanford, reached completion and operation, the only one of five to do so. Organizers of this initiative demonstrated in depth knowledge of the complex financial and legal issues on which nuclear projects depended. Once again, citizens asserted the right to participate in nuclear decisions.

Restart of PUREX

Neither Initiative 394 nor a state legislative investigation that preceded it touched on Hanford's military operations. These operations came under public scrutiny in the 1980s after President Jimmy Carter signed the 1980 Nuclear Weapons Stockpile Memorandum, which signaled to the United States Department of Energy (USDOE) that by 1988 it would need more weapons grade plutonium. The Reagan administration's 1982 request for money to produce 17,000 nuclear weapons over 15 years secured Hanford's future (Connor, 1985). Securing Hanford's future also invited public attention.

As part of the nuclear weapons stockpile buildup, the USDOE announced the restart of PUREX, the plutonium-uranium extraction plant at Hanford. While the restart of PUREX was a lightening rod for people opposed to nuclear weapons, it also raised concerns about the safety of PUREX operations. The Multnomah County Commission in Oregon, the Southwest Washington Board of Health and several citizen groups and peace coalitions asked that PUREX not be restarted until independent state oversight established the plant's safety. The USDOE denied the requests on the grounds that they came after the deadline for public comment (Shook & Connor, 1984).

A few months after the restart of PUREX, the *Seattle Times* ("Radioactive Thorium," 2/7/84) reported that the PUREX plant experienced a shutdown in operations two weeks earlier when radiation monitors detected a large amount of radioactive thorium leaking from a PUREX stack. The article quoted a Hanford spokesperson:

> We have accidents happening over here all the time . . . this emission produced no fatalities or serious injuries; there was no significant contamination of people or property. There was no shutdown of critical operations or facilities. So we didn't announce it.

Prior to and including the disclosure of the PUREX emissions, Hanford officials assured the public that any tank leak, air emission, or discharge into the Columbia River posed no hazard to the public. The public never challenged that claim—until the PUREX incident. In 1984, the Hanford Oversight Committee (HOC), a citizen action group, filed a Freedom of Information Act request for documentation of the PUREX release and

obtained documents they did not specifically request, including computerized logs of emissions from PUREX.

The HOC activists contacted Hanford activists in Spokane who had an
independent scientist, Dr. Allen Benson, analyze the computerized logs.
Benson determined the emissions from PUREX were thorium, which the
USDOE reported, and plutonium, which the USDOE had already denied.
During a meeting with Benson, Hanford scientists reversed their position
and conceded that PUREX had released plutonium the previous January.
Benson received a letter from Rockwell-Hanford officials who complimented
him for doing "an excellent job" of analyzing the company's documents
(Shook & Connor, 1984).

Several months later Rockwell-Hanford released a report admitting that
PUREX had released plutonium. One activist commented:

> It was Allen's work that, I think, probably produced the first major
> public retraction that Hanford ever made on an important techni-
> cal issue. They admitted that, yes, there was too much plutonium
> going up the stack of the PUREX plant.

Benson's work for the activists exemplifies the social movement model in which
citizens rely on independent experts to provide scientific credibility. Benson
eventually became affiliated with a group that formed in the fall of 1984, the
Hanford Education Action League (HEAL). HEAL later came to represent the
type of citizen action in which citizens become experts themselves.

Following the PUREX incident, investigative journalists Larry Shook
and Tim Connor wrote a four-part series on the safety of PUREX. As part of
their research, the journalists asked for the PUREX Environmental Impact
Statement (EIS) which the National Environmental Policy Act of 1969
required to be a public document. One of the journalists explained:

> We went to them and said, "May we have all this documentation?"
> And they wouldn't give it to us. It was an incredible run around.
> They said much of it was classified and we said, "Well this isn't an
> EIS. An EIS can't be based on classified information. . . .?" It was
> clear the EIS was just a part of the window dressing activity that
> they wanted to go through to assure the public that everything was
> okay.

The journalists were not assured. They compared the PUREX EIS to
another document known as ERDA 1538, which supposedly recorded all the
radiation ever released from Hanford. The two reports could not be recon-
ciled. A 15 curie release of plutonium from PUREX noted in the EIS did not
appear in ERDA 1538. Unwilling to accept USDOE claims that Hanford never
released enough radiation to harm the public, the journalists asked Hanford

officials for the documentation behind both reports. These requests were the first of many for documents that contributed to the release of the Hanford Historical Documents nearly two years later.

Hanford Education Action League

Many requests to the USDOE for documents came from the Hanford Education Action League (HEAL), a Spokane-based group. Founded in 1984, the organization formed in response to continued government secrecy about Hanford operations and the lack of democratic process in decisions about nuclear programs. HEAL wanted proof as to whether or not past and current Hanford operations posed a threat to public health and the environment. HEAL also called for the reestablishment of democracy at Hanford.

The experience of many of the founding members of HEAL was similar to the slow motion process of the emergence of public participation in decisions at Hanford. Many HEAL members were accepting of Hanford even though most had concerns about nuclear power technology. Concern about nuclear power or nuclear waste, however, typically led HEAL activists to Hanford. HEAL members moved from the first stage of acceptance and awareness to a stage in which self-education dominated.

Environmental activists are often charged with being "antiscience" when opposing technologies they consider hazardous (Dickson, 1988). In contrast, HEAL members became well educated on the scientific and technical issues of radiation health effects, Hanford operations, and nuclear waste storage. Self-education makes individuals less dependent on government officials and scientists for interpretation of information. The citizens become the experts. The HEAL activists came to perceive Hanford as a threat to public health and the environment and grew distrustful of the federal government. Concluding that the government had violated its promise to protect its citizens from harm, HEAL members believed the USDOE deliberately hid information from the public.

Although the process of self-education was ongoing, HEAL moved into collective action. It used every opportunity to ask the USDOE to prove the safety of Hanford operations past, present, and future. HEAL gathered and disseminated information about ionizing radiation and scientific disputes regarding low doses of radiation, Hanford operations, and the nuclear industry. The group published a newsletter and reports and held public educational meetings with scientists and government officials.

HEAL used a democratic approach to technical decision-making that pitted it against the USDOE's technocratic approach. HEAL took the position that citizens could be involved in evaluating Hanford's safety. The USDOE took the position that experts should decide.

Between HEAL's formation in 1984 and the release of the Hanford Historical Documents in 1986, HEAL helped shift the balance of power between the public and Hanford officials. HEAL refused to accept the government's conclusions about radiation health effects without reviewing the data and the studies themselves. HEAL provided testimony at public hearings and became a source of information for journalists when Hanford came under consideration as a location for a high-level nuclear waste repository. The government found itself unprepared for this type of sophisticated and informed citizen action. HEAL and other stakeholders pressured the USDOE to release documents, and eventually the agency ceded to these demands.

The Repository Debate

Wartime pressure initially excused government officials from dealing with the issue of permanent storage of nuclear waste. High-level waste entered temporary storage pending a long-term solution to the problem. Commercial nuclear power plants avoided the issue by using Hanford's interim storage concept until a repository for "permanent" disposal could be constructed.

In 1970, the Atomic Energy Commission (AEC) announced a commercial high-level nuclear waste policy that included construction of a national repository in a salt mine in Lyons, Kansas. A year later, the Lyons site proved an embarrassment to the AEC when testing resulted in water leakage into the mine (Carter, 1987). The AEC renewed the search for a suitable location by evaluating the technical suitability of multiple locations, including Hanford.

In the early years of site characterization, the repository issue received little attention. In the summer of 1979, the USDOE held its first public meeting in Seattle and informed the public of $30 million in expenditures on preliminary site characterization studies (Nalder and Connelly, 1979). Few Washington State legislators knew much about Hanford operations when the Nuclear Waste Policy Act (NWPA) of 1982 authorized the construction of a high-level waste repository. In the winter of 1983, the Washington legislature created the Nuclear Waste Board (NWB) with six agency directors, a chairperson appointed by the governor, and eight nonvoting legislators. The NWB provided oversight of the state's rights and interests in the repository siting. The governor also appointed a 15-member Nuclear Waste Advisory Council (NWAC).

The NWB attempted to assess baseline radioactive contamination at Hanford for environmental monitoring in the event that Hanford was selected as the repository site. This board found it as difficult to get documents from the USDOE as had journalists Shook and Connor. One NWB board member recalled that "Unbeknownst to most people in the state, there was a horse race going on in order to capture the dollars and get the repository." The Tri-City Nuclear Industrial Council wanted to see Hanford as a repository because it was confronting a bleak economic situation with two of the three planned WPPSS nuclear reactors canceled.

Both the NWAC and the legislature held public meetings across the state to inform the public of the repository issue. Testimony during these meetings changed the meaning of the repository from one of jobs and economic security for the Hanford area to a debate about safety. Stakeholders, galvanized by the issue, informed the public and politicians of studies that raised questions about the possibility of radiation leaking from the repository into the Columbia River and about potential dangers of transporting large quantities of radiation across the state.

Many environmental, public interest, and antinuclear groups united in opposition to the repository. Groups such as the Sierra Club, Greenpeace Northwest, HEAL and Physicians for Social Responsibility used various strategies to prevent Hanford's selection as the repository site. They conducted town meetings, distributed pamphlets and newsletters, and held forums as part of community education programs. Citizens used technical reports generated for the siting process to critique the process and to develop testimony for public hearings. Several groups used legal action to prevent the State of Washington from finalizing an agreement with the USDOE to define the state's role in operating the repository if USDOE selected Hanford as the location (Graydon, 1985). The City of Spokane adopted a resolution opposing shipment of waste to Hanford on Interstate 90, which passes numerous hospitals, high schools, and the business district.

Opposition to the repository intensified in 1985 when Hanford became one of three final candidates for the repository. Hundreds of people attended five USDOE hearings on the draft environmental assessment of Hanford as a repository site. The majority spoke in opposition to the selection although support continued to come from the Tri-City area.

While many people would cast the repository debate as one in which politics subverted science, the repository was an inherently political issue. The DOE, after decades of dictating nuclear policy, set out to site a repository purportedly on a technical basis. Repository opponents perceived the siting process as political and protested the power and authority the USDOE tried to use to achieve the outcome it wanted. This offered the media contrasting views to those of the USDOE and Hanford supporters. By highlighting flaws in the technical data that the USDOE used to justify its use of Hanford as a repository site, citizens effectively undermined the USDOE's authority. Thus the debate in Washington was both about the technology itself and the political relationships between the USDOE and various interest groups.

The Hanford Historical Documents

As stakeholders pressured the USDOE to release Hanford documents, newspaper accounts appeared describing concerns by some Hanford-area citizens. Known as downwinders, these people believed that exposure to radiation from Hanford caused illness and death among their families and friends (Steele, 1985). Following a 1985 conference sponsored by HEAL and the Spokane chapter of Physicians for Social Responsibility, Hanford's USDOE manager Mike Lawrence agreed to release Hanford documents in the interest of proving Hanford's safety.

On February 27, 1986, the USDOE released the Hanford Historical Documents with Mike Lawrence making the statement: "Our environmental monitoring indicates that there should be no observable health effects resulting from any of the releases that have occurred at Hanford in its forty-plus years of operations" (quoted in Connor, 1986). A week later, newspapers reported that the documents revealed a 1949 planned release of an estimated 5,500 curies of radioactive iodine. During a briefing of Northwest members of Congress, the USDOE acknowledged that Hanford released an estimated one-half million curies of radioactive iodine between 1945 and 1946 (Steele, 1986).

For months state officials, journalists, and citizen activists scoured the documents and produced a steady stream of revelations about Hanford emissions. As a result of these revelations and through its own review work, a panel convened by the governors of Washington and Oregon recommended that two studies be undertaken, one a dose reconstruction study to estimate the actual releases from Hanford and another to study possible health effects on the thyroid glands of people exposed to the atmospheric release of radioactive materials from Hanford (Ruttenber & Mooney, 1987). These two studies were funded and are nearing completion. They and two other government sponsored activities demonstrate other ways in which citizens participated in decision-making about technical matters.

The Hanford Environmental Dose Reconstruction Project

The Hanford Environmental Dose Reconstruction (HEDR) Project began in 1988 to estimate radiation doses the public may have received from the release of radioactive materials by the Hanford Site from 1944 to 1972. Until December 1995, a group of independent scientists, state and tribal representatives, and a public member, known as the Technical Steering Panel (TSP), directed the project. The completion of the study is currently being overseen by the HEDR Task Completion Working Group.

This arrangement was new for both the public and the scientists at Battelle Pacific Northwest Laboratories who were under contract to perform the technical work of the HEDR project. The scientists had never had their work directed by an independent group such as the TSP. Nor had members of the TSP previously encountered such public scrutiny as they garnered during the HEDR project. As a Hanford contractor, Battelle had no credibility with the public. As the technical directors of the HEDR project, the TSP had to earn the public's respect, a challenging feat given the USDOE as its funding source and the use of Battelle as the contractor (Niles, 1996).

At the TSP's fourth meeting in November 1988, Jim Thomas, the research director for HEAL, raised concerns about credibility. He stated that the TSP had failed to live up to its promise of an open study. Judith Jurji, a Hanford downwinder, reminded the TSP that "There must be no secrecy, no classified information, no barring of the press or citizens groups. And when data becomes available it must be made public promptly" (quoted in Niles, 1996, p. 16).

The members of the TSP, however, had established some policies and procedures that diminished the public's access. In the interest of pursuing science in the usual way, the TSP planned work sessions that would be used to discuss scientific results with peers. When a member of the public wanted to attend a work group closed to the public, TSP chair John Till reversed the policy. He recognized that the HEDR project was not a scientific study but a public study. Till reflected:

> As a result of all this, I want to tell you that I did some very hard soul searching. I have changed my attitude about this, and how I think we should be thinking in terms of openness in the future.... There is absolutely nothing that I have ever done, ever thought, notes I have ever written, nothing whatsoever, why a member of the public, if they want to be there to check me and talk to me about, could not be there. (quoted in Niles, 1996, p. 19)

As a result, public involvement was a large component of the panel's work. All meetings were open to the public, public comment periods were available during the business meetings and criticisms and concerns of the public were responded to by the panel members or its contractors. While the TSP accomplished something new, conducting science in the public domain, it also experienced intense public scrutiny. Groups like HEAL served as watchdogs and provided critical information to the public.

At the end of phase one of the project, HEAL pointed out that there were significant deficiencies in the work of the HEDR Project. One example of the deficiencies was that although the TSP acknowledged more work was needed to evaluate the significance of plutonium and ruthenium particle releases, the work plan downplayed the seriousness of the problem suggested by historical documents. HEAL appealed to the Centers for Disease Control and Prevention (CDC), the federal agency that manages the HEDR Project. HEAL asked for and received a more aggressive review of this particle problem.

Hanford Thyroid Disease Study

The second of the two studies recommended by the panel was the Hanford Thyroid Disease Study (HTDS). The study was mandated by Congress in 1988 to investigate whether or not there is an increased incidence of thyroid disease among people exposed to the atmospheric releases of radioactive materials from Hanford between 1944 and 1957. The protocol used for the study underwent extensive professional peer and public review.

One concern raised by members of the scientific community and the public was that the study did not include use of thyroid ultrasound as part of the initial examination of study participants. One Hanford activist met with the principle investigators of the HTDS and was dismissed as being unqualified to question their decision not to use the procedure. Disturbed but undaunted, the activist contacted the Centers for Disease Control and Prevention (CDC), which funds the study. Since other reviewers had raised the same issue, the CDC convened a panel of experts to discuss the matter. While the meeting did not result in a formal recommendation, the HTDS Management Team presented a proposal to CDC to include thyroid ultrasound exams for all study participants. In September 1992, the CDC's Advisory Committee voted unanimously to include thyroid ultrasound scans in the HTDS (Hanford Thyroid Disease Study, 1992).

Hanford Health Effects Subcommittee

A final example of current citizen participation in decisions about Hanford is the Hanford Health Effects Subcommittee (HHES). This is a federally chartered group that advises the Agency for Toxic Substances and Disease Registry (ATSDR) and the CDC about studies and public health activities related to Hanford. The subcommittee meets several times a year throughout the Northwest. The members of the Subcommittee represent different stakeholders, including labor, downwinders, health care, science, environmental groups and the nuclear industry. The diverse range of opinions on the panel provides for stimulating, sometimes contentious, discussions, but ultimately citizens are responsible for participating in the development of scientific projects.

Since its inception in 1995 the Subcommittee has reviewed and endorsed an ATSDR study on Hanford Fetal Deaths and Infant Mortality Analysis for the years 1940–1952. The HHES also recommended and assisted ATSDR with the development of a medical monitoring program and subregistry for people exposed to iodine-131 from Hanford. Although the USDOE agreed to fund the medical monitoring project with an initial $5 million of the projected $13 million cost, there has not been a release of funds.

In the spring of 1998, Hanford downwinder Trisha Pritikin, who is also a member of the HHES, sued the USDOE seeking to compel the agency to fund the medical monitoring program. Pritikin sued under the Superfund law that governs Hanford's site cleanup. The suit was dismissed by Judge Edward Shea who agreed with the government's contention that the Department of Energy has sovereign immunity and cannot be sued to force payment of nearly $13 million for medical monitoring sought by some 14,000 downwinders. Pritikin filed an appeal of the dismissal in U.S. District Court in the spring of 1999.

Conclusion

The emergence of public participation in decisions at Hanford was a process that developed in slow motion. Hanford was like a chronic disaster that:

> gathers force slowly and insidiously, creeping around one's defenses rather than smashing through them. The person is unable to mobilize his normal defenses against the threat, sometimes because he has elected consciously or unconsciously to ignore it, sometimes because he has been misinformed about it, and sometimes because he cannot do anything to avoid it in any case. (Erickson, 1976, p. 255)

As the cloak of secrecy surrounding Hanford operations unraveled, public pressure to participate in Hanford-related decisions increased. The shift in perception of Hanford from a symbol of military might and progress to one of a risk empowered citizens to collective action to oppose policies that suppressed and withheld information about Hanford.

In this case study of Hanford, citizens demonstrated the ability to take an active role in deciding what science and technology policies pose a danger to public health and the environment and the ability to work to change those policies. The government brought citizens and experts together to formulate policies that fused social values and technical data. Such policies are likely to serve the people most affected by them better than policies development without citizen input. Public participation and oversight in the policy process, either through institutionalized methods or collective action, is likely to enhance public acceptance of "public" policies.

Bibliography

Advance Advertising. (1983). *Alive! Yesterday and today: A history of Richland and the Hanford project*. Richland, WA.

Bigelow, J. (1950, January 29). AEC spending half billion to rebuild plutonium plant. *The Seattle Times*, p. 33.

Boyer, P. (1985). *By the bomb's early light: American thought and culture at the dawn of the atomic age*. New York: Pantheon Books.

Brown, P. (1987, summer/fall). Popular epidemiology: Community response to toxic waste-induced disease in Woburn, Massachusetts. *Science, Technology, and Human Values, 12*, 78–85.

Carter, L. J. (1987). *Nuclear imperatives and public trust: Dealing with radioactive waste*. Washington, DC: Resources for the Future.

Caufield, C. (1989). *Multiple exposures: Chronicles of the radiation age*. Chicago: University of Chicago Press.

Connor, T. (1985, winter). Does America need plutonium? *Pan Environmental Journal, 1*, 18–34.

Connor, T. (1986, March/April). Hanford Documents. *HEAL Newsletter.*

Couch, S. R., & Kroll-Smith, S. (1997 October). Environmental movements and expert knowledge: Evidence for a new populism. *International Journal of Contemporary Sociology*. Vol. 34, No. 2: 185–210.

The Daily Olympian. (1998, October 8).

Dickson, D. (1988). *The new politics of science*. Chicago: The University of Chicago Press.

Erickson, K. (1976). *Everything in its path: Destruction of community in the Buffalo Creek flood*. New York: Simon and Schuster.

Gamson, W. (1988). Political discourse and collective action. In *International Social Movement Research*, vol. I (pp. 219–244). Greenwich, CT: JAI Press. Editors: Klandermans, B., Hanspeter, K. & Tarrow, S.

Gilje, S. (1971, January 22). 5000-job cutback feared at Hanford. *The Seattle Times*, p. 1.

Graydon, D. (1985, January 10). Five groups sue over N-dump plans. *The Seattle Post-Intelligencer*, p. A6.

Hacker, B. (1987). *The dragon's tail: Radiation safety in the Manhattan project, 1942–1946*. Berkeley: University of California Press.

Hanford will stay secret, says Truman. (1946, March 8). *The Seattle Times*, p. 1.

HTDS Newsletter, 2(3), (1992, September).

Krimsky, S., & Plough, A. (1988). *Environmental hazards: Communicating risks as a social process*. Dover, MA: Auburn House Publishing Co.

Lane, B. (1976, February 3). N-plant construction block looms. *The Seattle Times*, D13.

Lane, B. (1980, October 21). A 'friendly' nuclear ballot issue? *The Seattle Times*, p. A8.

Levine, A. G. (1982). *Love canal: Science, politics and people*. Lexington, MA: Lexington Books.

Lilienthal, D. (1971). *The journals of David Lilienthal: Volume V—The harvest years 1959–1963*. New York: Harper and Row.

Niles, K. (1996). *Reconstructing Hanford's past releases of radioactive materials: The history of the Technical Steering Panel 1988–1995*. Prepared with support from the U.S. Centers for Disease Control and Prevention.

Nalder, E., & Connelly, J. (1979, August 9). Questions over N waste. *The Seattle Post-Intelligencer*, p. A8.

Norman, C. (1984, January 20). High level politics over low level waste. *Science, 223,* 258–260.

Radioactive thorium leak at plutonium plant no cause for alarm. (1984, February 7). *The Seattle Times*.

Read, T. (1975, January 19). Radioactive leaks plague Hanford. *The Seattle Post-Intelligencer*.

Ruttenber, J., & Mooney, R. R. (Eds.). (1987). *Report of the Hanford Health Effects Review Panel and recommendations of sponsoring agencies*.

Sanger, S. L., & Mull, R. (1989). *Hanford and the bomb: An oral history of World War II*. Seattle: Living History Press.

The Seattle Times. (1971, January 22). *The Seattle Times*.

Shook, L., & Connor, T. (1984, September 3). Purex plant raises safety questions. *Bellevue (WA) Journal-American*, pp. A1, A3.

Smyth, H. D. W. (1945). *Atomic energy for military purposes: The official report on the development of the atomic bomb under the auspices of the United States government, 1940–1945*. Princeton: Princeton University Press.

Steele, K. D. (1985, July 1). "Downwinders"—Living with fear." *Spokane (WA) Spokesman-Review*, p. 1.

Steele, K. D. (1986, March 6). In 1979 study, Hanford allowed radioactive iodine into area air. *Spokane (WA) Spokesman-Review*, p. A6.

Technical Steering Panel of the Hanford Environmental Dose Reconstruction Project. (1994). Representative Hanford Radiation Dose Estimates.

Unsoeld, K. (1983, January). Should public power be nuclear power? *Dollars and Sense*, 15–17.

Weart, S. (1988). *Nuclear fear: A history of images*. Cambridge, MA: Harvard University Press.

Whitely, P. (1981, June 26). Court throws out ban on N-waste. *The Seattle Times*, p. A1.

Part II

Assessments and Strategies

5

Human Well-being and Federal Science

What's the Connection?

DANIEL SAREWITZ

This brief essay considers how the organizational roots of federally funded science influence the capacity of the nation's research enterprise to contribute to human well-being. I take well-being (individual and collective) to have several components, including: (1) the fulfillment of all elemental needs necessary for survival; (2) the achievement and preservation of human dignity; and (3) the capacity to act on a more or less level civic and moral playing field.[1] The essence of my argument is that the Cold War origins of the federal research enterprise, and the philosophical foundations on which this enterprise rests, are implicated in a range of tensions and challenges to human well-being that the enterprise–as currently organized—cannot to address coherently. These tensions arise in large part from the fact that science aims at delivering benefits to society through the achievement of predictive certainty and technological control, while the vitality of both nature and democracy derive from a lack of predictability and controllability. My discussion starts with the organization of federal (U.S.) science, but considers human well-being in a global context. This connection is reasonable because the U.S. is by far the most scientifically productive nation, and because the impacts of science on society are global in character.

Cold War Roots

The role of the military in organizing the nation's current science and technology enterprise cannot be overstated. From the end of World War II until the launch of Sputnik by the Soviet Union in 1957, 80% or more of all federally funded research was justified in terms of national security needs. The creation of the American research university and the explosion of technology-intensive industries that lay at the core of the nation's economic growth were strongly and directly catalyzed by funding from the Department of Defense. Moreover, when Sputnik stimulated a highly politicized call for an increased national commitment to civilian research, the lion's share of resources during the subsequent decade went to the manned space program, which in many ways was simply a technological adjunct to the Cold War defense effort. For example, many of the information management, advanced materials, and navigation and control technologies necessary for space travel were also applicable to—or borrowed from—the nation's high-technology defense system.[2]

In the early 1950s, as the scale and complexity of America's Cold War geopolitical commitments became clear, research and development came increasingly to be viewed in the Defense Department and among leading science administrators not simply as the provider of particular weapons systems, but as a continual source of new knowledge, innovation, and technical expertise that would preserve American military preeminence across a diverse range of potential national defense applications. The scientific-military nexus entrained—and sustained—all sectors of the post-War research enterprise. From the perspective of those who designed and built this enterprise, the important functional distinction in science was not between basic and applied, but between classified and unclassified. The knowledge-production process was viewed not in terms of particular disciplines of basic science, but specified outcomes for military needs. The university was viewed not as an ivory tower, but as a vital cog in the national defense machine that included private industry and a range of government agencies. The role of the scientist was not as maverick roaming the frontier of knowledge, but as an interactive member of a multitalented research group that often included theorists, experimentalists, and engineers. While research tools such as particle accelerators were used to carry out what might be termed *pure* research on fundamental physics, they were paid for by the Defense Department and the Atomic Energy Commission in large part because they were valuable test beds for technologies with military applications, and because they were the training ground for scientists who would help to create the coming generation of Cold War weaponry.[3]

The research organization that flowered from these Cold War roots was thus dominated by physical science, justified in terms of its role in technology development, and characterized by a dependency relationship between scientists, be they governmental or not, and their sponsoring federal agencies. The persistence of this organization can be seen in the continued dominance of three agencies—the Department of Defense, NASA, and the succession of energy research agencies—which peaked at nearly 90% of the federal R&D budget at the height of the Apollo program in 1965, and today still constitutes 66% of all federal research and development spending. New political momentum to develop missile defense systems, at expenditure levels that dwarf most other research programs, demonstrates that these patterns will continue for the foreseeable future.[4] Even in academia, many important fields, such as electrical engineering, computer science, and materials science, are today strongly supported by Defense Department funds, while physics continues to derive much of its support from the Department of Energy, which is a direct descendent of the Atomic Energy Commission (Wulf, 1998, p. 1803).

One consequence of this organizational heritage is that the tools at hand are applied to emerging issues and problems, even if they are arguably inappropriate. For example, the massive U.S. Global Climate Change Research Program (USGCRP), which was established in law in 1990, the year after the Berlin Wall fell, is dominated by NASA space technology programs for data acquisition, and physical science approaches to modeling and interpreting atmospheric (and, to a lesser extent, oceanic) processes and evolution (cf. Subcommittee on Global Change Research, 1997). One could easily imagine an alternative (and less expensive) program that placed a considerably greater emphasis on the life and social sciences—indeed, a growing sentiment for such a reprioritization is now beginning to emerge in some quarters—aimed at understanding the dynamics of ecosystems and social systems in a changing global and policy environment (cf. Lawler, 1998). The organizational and political basis for such an effort, however, was not in place at the time the USGCRP was being planned. Our national approach to climate change thus strongly reflects the organization of Cold War science.

While the Cold War justified a top-down approach to setting science priorities, the ideology of basic research called for a bottom-up arrangement where scientists themselves would determine the most fruitful directions for fundamental investigation, based on their expert judgment as exercised through peer review and other mechanisms. This ideology has been most successfully implemented through research funded by the National Science Foundation and the National Institutes of Health. To a very great extent, of course, the ideal of an autonomous scientist exploring the frontiers of nature is strongly buffered by bureaucratic and policy decisions about how and where to allocate money, by the organization of research institutions such as universities and federal laboratories, and by the disciplinary organization of science itself, but within these constraints there is no question that the state of fundamental knowledge about nature has been spectacularly advanced by federally funded scientists acting with individual autonomy. All the same, individual autonomy can be exercised–and science advanced—in settings that are severely bounded. For example, during the Cold War, the conduct of classified military research on universities campuses was commonly justified by the argument that scientists and engineers had to be free to pursue whatever research they chose, regardless of whether or not it was subject to the strictures of military secrecy (Lowen, 1997, p. 144). An even more extreme case is illustrated by the Soviet Union under Stalin, where scientists somehow managed to conduct fruitful and sometimes world-class fundamental research programs, even as they were subjected to severe political persecution that often included prison and torture (Graham, 1998).

The Enlightenment Program

Given this combination of top-down organization motivated by the Cold War, and bottom-up research trajectories determined in part by individual scientists, how is human well-being introduced into the equation? After all, the promise of science to fulfill human needs is perhaps the principal political justification for public funding of civilian research. Over much of the past 50 years, this question has often been sidestepped by the assumption and assertion that the benefits of science flow automatically to society, deriving spontaneously from the progressive increase in the reservoir of fundamental knowledge, the development of innovative and beneficial new technologies, and the operation of the market economy that introduces the products of science and technology into society. In furthering this view, one can reasonably argue that the *program* for science (to use a term favored by social scientists) that was articulated and embraced by great Enlightenment thinkers from Bacon and Descartes to Jefferson and Voltaire, and reframed in a modern guise by Vannevar Bush's famous report, *Science, The Endless Frontier* (1960/1945), has been almost inconceivably successful. This program prescribed the linking of scientific knowledge about the laws of nature to the technological control of nature itself for the benefit and progress of humanity; it was implemented in its most comprehensive and successful form by the Cold War organization of American science; and it is internalized today at every level of the diverse and complex modern research enterprise, and throughout industrialized society as a whole. One can choose one's symbol of the culmination of this program—the polio vaccine; the hydrogen bomb; the invention of the transistor; the cloning of Dolly the sheep or the defeat of world chess champion Garry Kasparov by Deep Blue the computer—but the overall point seems unavoidable: human affairs are now mediated at every level and on every scale by science-based technology and the economic activity that it engenders. There are optimists who view this achievement as the salvation of the species (and even of nature [Ausubel, 1996]) and pessimists who portray it as a disaster-in-progress, but few if any would argue with the observation itself.

Neither would many suggest that we have arrived at utopia. But the idea that some desirable type of steady-state if metastable condition for humanity may actually be within reach has in fact taken hold of optimists and pessimists alike, as embodied in the environmental and economic concept of sustainability, the political ideal of "the end of history," and the technologist vision of nature as infinite cornucopia and the human life span as infinitely extendable. That such ambitions can be taken seriously and exist simultaneously in our jaded and hypersophisticated age is strong confirmation that the Enlightenment program for science has fulfilled its ambitions.

Yet the reality of both nature and democratic governance reveal crucial tensions embodied in the Enlightenment program for science. In terms of nature, the central paradox is that while the scale of control afforded by science and technology continues to increase, so does the domain of uncertainty and potential risk. "Knowledge of the [natural] system that we deal with is always incomplete," writes the ecologist C. S. Holling. "Surprise is inevitable . . . there is an inherent unknowability, as well as unpredictability. . . . The essential point is that evolving systems require policies and actions that not only satisfy social objectives [i.e., control for human benefit] but, at the same time, also achieve continually modified understanding of the evolving conditions and provide flexibility for adaptation and surprises" (Holling, 1995, pp. 13–14). Whereas science-based technological control of nature for human benefit has been the hallmark of the Enlightenment program, this control has aimed at severing the ties between society and nature, of escaping the threats and uncertainties imposed by those ties, of overcoming the natural limits on human endeavor in every activity, from food production to cognition. Holling's insight is that this goal—freedom from natural caprice—is unachievable because the very act of controlling natural systems introduces new variables that increase the unpredictability of the systems' dynamics. This insight has been borne out repeatedly in failed efforts to "manage" ecosystems. Protecting and preserving complex natural systems—systems on which human survival depends—thus requires that the expectation of control be abandoned, and replaced by an awareness that we cannot dictate the consequences of our actions in nature. The completely unanticipated discovery of the Antarctic ozone hole in 1985 is a stark example of this problem. In exercising control over the local environment through the use of chlorofluorocarbon refrigerants, we also perturbed the component of the stratosphere that protects the earth's surface from ultraviolet radiation. As the scale of human activity increases, we should expect surprises such as this to become more common.

Less recognized and appreciated is a similar relation between democracy and the Enlightenment program for science. It is often said that the price of democracy is "eternal vigilance." This means that democratic society must constantly guard against monopoly in the competition among ideas, principles, and morals. The competition itself is crucial to the existence of democracy. In its absence is oligarchy and authoritarianism. Human civilization has been marked by an ongoing discourse about the essential attributes that characterize a "good" society and its citizens—freedom, justice, equality, wisdom, mercy, tolerance, restraint, sharing—but there will never be an equilibrium equation describing the perfect, utopian balance among these attributes. Conflicts are inevitable and necessary, trade-offs must constantly be made—justice must be *tempered* with mercy; freedom with restraint. This struggle proceeds through a succession of social and political consensuses that must be hashed out in a process that is reasoned but not strictly rational, practical but not pretty. In a civil society, "there can be no ultimate closure," writes the social theorist Philip Selznick, "because values reflect existential conditions, which are always subject to change. . . . Reason takes into account the temptations and limitations of human conduct; therefore it is self-critical and self-limiting. This moderating outcome is also a source of indeterminacy. . . . Certainty is sacrificed on the altar of reason" (1992, p. 61–62).

The fact that indeterminacy is not only inevitable but essential to democracy—something to be embraced rather than overcome—does not comport well with a scientific worldview whose most legitimating measures of success are predictive certainty and control of nature. Having created the material welfare and technological infrastructure on which democracy has now come to depend, the significance of Enlightenment science for the democratic process itself seems murky. The issue here is not only that science and technology constantly transform the structure of society (usually without the consent of the governed), but also that over the past half-century or so, they have become tools that are called on to assist in the explicit improvement of democratic process—helping to resolve political dispute, set priorities for action, and manage social change. Through scientific eyes the mechanisms that enable democracy—politics, laws, bureaucracy—may look not just messy and irrational but subject to scientific correction. But the quest for scientifically based certainty and control in the political realm can conflict with democratic ideals by demanding, and fueling a demand for, that which is fundamentally incompatible with civil society: closure.

Eight Problems

The scientific ideals of precision, determinacy, and control may thus lie in profound tension with society's struggle to comprehend and manage complex and evolving systems, be they natural or democratic. It can be no surprise, then, that while the Enlightenment program for science, catalyzed by an organizational structure that grew out of the Cold War, has helped to create heretofore unimaginable breadth of knowledge, degrees of control and levels of affluence, it has also revealed and provoked new challenges, contradictions and conflicts. Perhaps most prominent among these is that nature has gone global. Issues such as climate variability (El Niño, global warming), disease migration (AIDS, ebola), and invasive species (kudzu, killer bees) reflect the interaction of a global industrialized economy and a closed but infinitely complex earth system. The range of direct if uncertain threats to human well-being is daunting. These threats demonstrate that a science and technology program focused on ceding control over the environment to individuals acting quasi-independently yields unanticipated complications at higher organizational levels of nature and society. Ozone depletion has already been mentioned as one such complication. Continued sea level rise is another, with impacts exacerbated by the ongoing migration of populations to coastal regions. The cumulative effects of global ecosystem destruction loom as a threat of unknown but potentially disastrous proportions. And such problems are greatly complicated by fundamental issues of justice and equity. Obviously, the adverse impacts of global environmental problems are experienced disproportionately by poor people and poor nations, who have fewer resources and less flexibility in responding to changing environmental conditions (cf. Sachs, 1996). Science is not organized to integrate such considerations of equity into its research priorities.

Yet even as these types of disparities grow more severe, a second problem can be recognized. As the affluence of industrialized nations of the world continues to grow, the priorities and capabilities of science and technology (fueled by the market incentives that promote commercial application of science) become increasingly divorced from the basic needs of those people—in rich nations as well as poor—who have not proportionately benefited from the products of the Enlightenment program. The clearest example is biomedical research, which in the U.S. focuses its formidable resources on fundamental investigations into molecular function and the search for high technology cures for diseases of affluence and old age. These priorities largely neglect a broad range of low-cost (and low-profit) opportunities in public health, such as nutrition research; they also fail to meaningfully address the debilitating health care problems of the developing world (cf. World Health Report, 1996). Even when a problem is of a global scale, the benefits of science may be disproportionately appropriated by affluent societies. This moral dynamic is strikingly illustrated in the case of AIDS, where high-technology and high-priced drug therapies are having a remarkable measure of success in reducing mortality among AIDS-sufferers who live in developed nations. In the developing world, the story is entirely different, and in many nations where advanced drug therapies are unaffordable, AIDS has become not only an overwhelming public health problem, but a tangible threat to economic and social prospects as well (Warrick, 1998, p. A2). Understanding and controlling the HIV virus at the molecular level is the very stuff of the Enlightenment Program: high-prestige science that attracts money, peer approval, and Nobel prizes. Again, the organization that encourages and rewards such science includes no provision to change research trajectories in response to a moral imperative.

Third, rapid progress in science and technology is transforming the fundamental institutions of civil society—including the structure of community, and the democratic process itself—in ways that are neither well understood nor easily controlled. Moreover, the speed of scientific and technological progress is such that profound and often wrenching societal transformations appear with greater frequency than ever before in history. In recent years, the erosion of civic community in America has been a subject of much concern to intellectuals, politicians, and pundits of every stripe. While the causes of any such decline must be complex, the proliferation of one transforming technology after another—telegraph, telephone, automobile, radio, television, air conditioner, home computer, Internet—introduces a pervasive instability into the structure and function of community. Similarly, within the period of a generation or two, the Green Revolution made the idea of the small, family-owned farm both economically and technologically obsolete (in the absence of government subsidy, at least) across much of the globe. The Enlightenment Program supports a social consensus that accepts such change as the unavoidable price of progress. The Program simply does not accommodate the idea that criteria of societal well-being are a legitimate guide for and metric of scientific progress.

Fourth, an overlay on this transformation process seems to be that technological progress exacerbates inequitable distribution of wealth, both within and between nations. A nation with many scientists and engineers, many telephones, computers, universities, and high technology-companies will generate more ideas, more opportunities, more productivity and economic growth, than a nation that lacks these assets. This kind of growth is self-perpetuating, and has resulted in, among other things, a considerable increase in concentration of wealth among the world's industrialized nations over the past three decades—despite the remarkable gains made by a few East Asian nations (cf. United Nations Development Programme, 1992). This phenomena contradicts a basic tenet of the Enlightenment Program—that new knowledge is cosmopolitan in its benefits—and demonstrates that scientific knowledge and technological control are appropriable commodities. Indeed, the Cold War organization of science was in its essence a quest to generate national advantage through the appropriation of knowledge and innovation.

Fifth, scientific uncertainty has become an increasingly common cause of political gridlock, especially in controversies related to the environment and natural resources. Global climate change is the archetypical example of this trend. The technical debate over the scientific validity of global warming has become a surrogate for a value debate about the preservation of the environment and the distribution of the benefits of industrialization. A sixth, possibly related issue is the overwhelming increase in the volume and availability of technical information relevant to human decision making at every level of society, unaccompanied by indications that this trend is in fact leading to greater wisdom or better decisions in the public sphere. These two trends reflect the Enlightenment confidence that more scientific information is in itself sufficient to drive political solutions to a range of societal problems. Scientific debate thus becomes a surrogate for underlying value debate necessary to make decisions, while the demand for "more information" displaces the demand for effective decision-making.

Seventh, and partly as a reflection of its own success, the research enterprise is increasingly caught up in vexing and divisive ethical questions, such as those surrounding the cloning of higher organisms, technological erosion of privacy, patenting of genetic material, and clinical testing of new medicines in different cultures. Such questions often reflect a collision of scientific progress, commercial incentive, and ethical norms. For example, personal privacy may be threatened by the availability of genetic information that could be used as a basis for offering or denying medical or life insurance coverage to individuals (Hudson, Rothenburg, Andrews, Kahn, & Collins, 1995, pp. 391–393). In cases such as this, ethical conflict is an inevitable byproduct of the success of the Enlightenment Program. The proliferation of new knowledge and techniques for the control of both nature and societal activity lie in profound tension with the desire of democratic society to exercise control through political processes. This conflict is exacerbated by the operation of the open market, which seeks to introduce the products of science into the economy, and resists any efforts to restrict its ability to do so.

Finally, the research community itself increasingly reports on a breakdown in confidence, optimism, and morale, especially in academia. Explanations of this phenomenon include loss of autonomy due to increased demand for scientific accountability to the public; ever-increasing degrees of specialization that alienate scientists from the real world; and the dissolution of community due to competition among peers for recognition and funding (cf. Pollack, 1997). These explanations suggest an increasing tension between the evolving role of scientists in society, and some key tenets of Enlightenment science, for example, that scientists are accountable to society only through the quality of their science; that the sure path to comprehending nature is reductionism; that individual productivity is the key to scientific progress.

Now it may not be unreasonable to pronounce the foregoing issues to be political, sociological, and economic in nature, and thus largely beyond the capacity of science to address. One might also simply observe that while social progress is often slow, and politics irrational and difficult, more science and technology—a continuation of the Enlightenment program—cannot help but overcome these hurdles and continue to move things in the right direction, as they have done in the past.

But what is crucial here is not simply that the building and shaping of the entire portfolio of federal science and technology activities in the early post-War years took place *without regard* to any of these problems (many of which, of course, either did not exist or were not recognized), but as well that the organizational structure and knowledge products of today's enterprise are often not suited to addressing them productively. While the explicit question of how best to connect the enterprise to human well-being has raised its head from time to time in political debate over how science and technology should be organized, the idea that such beneficial connections will arise automatically through implementation of the Enlightenment program has always, in the end, prevailed.

Thus, recent consideration of how federal science should be connected to human well-being is usually framed in terms of how to enhance the performance of the Enlightenment program, not in terms of questioning the program itself. For example, the science community and its many advocates focus strongly on the need for "better communication" between scientists and the public, in order to ensure continued public support for research funding. The ideal of the "civic scientist" or "citizen scientist" has begun to get some attention from such scientific leaders as the director of the National Science Foundation (and current Presidential Science Advisor) (Lane, 1996). While this ideal is certainly laudatory, as commonly articulated it is also rooted in the assumption that the feeding tube that delivers information from the scientific community to the public just needs to be widened, and it neglects the perhaps more fundamental issues of what type of nourishment, exactly, is being provided, and what the public might have to offer the research enterprise in terms of wisdom and guidance. "Sound science" is also commonly invoked as a cure for the political battles that often emerge over science- and technology-related issues such as global environmental change, biodiversity preservation, energy policy, and nuclear waste disposal. Technocratic approaches such as risk assessment are often prescribed to help rationalize the political decision making process (House Committee on Science, 1998). Again, these types of prescriptions, while not without some potential application, accept the received, Cold War organization of the science and technology enterprise as a starting point. I am arguing instead that the fundamental operational realities of democracy and nature, which dictate an essential unpredictability and uncontrollability, render the continued implementation of the Enlightenment Program through the mechanisms of the

Cold War inherently problematical. New organizational tenets and models may be necessary to confront these difficulties.

New Links Between Science and Well-Being

If the organization of science is understood to significantly reflect social, political, and economic processes—especially the very specific if multifaceted priorities dictated by the conflict between the United States and the Soviet Union—then the eight problems mentioned above can be seen as defining a reality within which new links between science and human needs can potentially be forged. The shape of some of these new links is gradually becoming apparent. New philosophical approaches now compete with the key insights of the great Enlightenment thinkers. The most conspicuous and controversial of these has arisen from the field of social studies of science. This approach sheds important light on the social and political context within which scientists are working and scientific problems are defined and confronted. But it has been reviled and rejected by most mainstream scientists—and not a few social scientists—for its assertion that scientific knowledge is socially constructed. In my view, this assertion is simply trivial as a critique of science per se: In the real world, the success and impact of science is argument enough for the validity of its method and philosophical underpinnings—socially and politically constructed as they may be. As David Hull writes: "No amount of debunking can detract from the fact that scientists do precisely what they claim to do" (1988, p. 31). This success, however, is necessarily and appropriately defined within the context of the Enlightenment program, and especially the application of scientific knowledge to the technological control of nature. A more fruitful question, then, would address the extent to which the Enlightenment program is appropriate for and compatible with the types of challenges facing society today. The social studies of science have helped to position science where it belongs—in the heart of society, rather than as an insular satellite—and even through the rancor it stimulates, brings attention to this question, and thus raises the possibility of alternative programs for the future.[5]

As a more practical matter, mechanisms to better connect democratic process to the establishment of scientific priorities and practices are fitfully beginning to develop, although often over the strenuous objections of the scientific community. In Europe, citizens conferences that give communities the opportunity to make decisions about scientific and technological choices are gradually becoming recognized as an effective tool, while in the United States, environmental stakeholder groups, often organized at the scale of a local watershed, now increasingly replace, augment, or subsume expert technical debate in the effort to resolve environmental dilemmas. The National Institutes of Health has begun placing patients on a few of its peer review panels to help introduce broader perspectives into the process of allocating biomedical research funds. And community-based research is a nascent but highly promising mechanism for linking scientists to local people and problems in a manner that was never envisioned or allowed for in the Cold War organization of science (cf. Sclove, 1995; Landy, Susman, & Knopman, 1999; Agnew, 1999; Sclove, Scammell, & Holland, 1998).

Efforts to achieve a more synthetic view of nature by breaching the barriers that separate traditional scientific disciplines are occurring with variable success in such disparate areas as cognitive neuroscience and environmental science. The emerging and well-publicized field of complexity science acknowledges the intrinsic limitations of traditional approaches to understanding and describing nonlinear systems such as consciousness or economies. Throughout the sciences, there is an increasing recognition that nature is simply not comprehensible in terms of narrow, disciplinary, reductionist investigation (cf. Cornwell, 1995).

The concept of sustainability is giving birth to new measures of progress and new agendas for research that explicitly link science to a moral framework rooted in the tenet of intergenerational equity. Sustainability has become a guide for research in diverse fields, such as agriculture, economics, ecology, and public policy. The idea of adaptive management of complex systems acknowledges that such systems (natural and social) are not predictable, and that both science and policy are thereby always subject to error and amenable to correction. Adaptive management recognizes that values are usually less malleable than science, and thus prescribes for science the role of assessing and monitoring the impacts of policy decisions that have already been made. Science becomes a tool for correcting and improving the incremental democratic policy process by providing insight, rather than dictating policy by providing predictions. Industrial ecology views manufacturing and energy use as cycles, rather than independent streams of production, consumption, and waste, and thus defines entirely new criteria for judging the viability of technologies (cf. Lee, 1993; Graedel & Allenby, 1995). Sustainability is a goal; adaptive management is a policy process for moving toward the goal; industrial ecology is a technical perspective that supports the policy process. These linked concepts view nature and democracy as models to be emulated and supported, not obstacles to be overcome.

Progress in these and related directions is important and promising, but the relevant scientific activities remain a marginally small proportion of the total federal science enterprise. Typically, it seems, initiatives along these lines are undertaken in isolation, often as a result of the action of individuals with vision and energy. There are few institutional structures within which successful experiments can grow and propagate. Funding sources and reward systems still militate against those who would seek to define stronger connections between science and human well-being.

One problem, of course, is that the organizational inertia of the Cold War is difficult to displace. Doing so will take time and persistence. Another problem is that the mental model of the Enlightenment program is so simple and elegant—more science and more control always yield more well-being, in essence—that it resists being replaced by something more nuanced. But a few things can be said about the components of an alternative organization for science that seeks to relieve the tensions created by the Enlightenment program. This new organization will certainly direct us toward new types of institutions that promote, as a foundation for determining research priorities, meaningful interaction between scientists and the people who scientists are serving. It will portray complex, real-world problems, rather than scientific disciplines, as organizing foci for research programs. Artificial taxonomies, such as basic versus applied research, will be abandoned as meaningless, and the boundaries between social and natural science will become increasingly permeable. Metrics of scientific excellence will focus as much on social outcomes as on scientific ones. Scientists and science administrators will internalize the idea that complex and indeterminate structures of nature and democracy must be a basis for such new organizational approaches to research as adaptive management and industrial ecology.

The transition to an organization of science characterized by these and related attributes is a matter, for the most part, of political vision and will. It is worth emphasizing that whereas the mythologies of the "golden age" of Cold War science tell a story of abundant funds available to individual scientists who freely pursued exciting new knowledge where ever it might lead, the broader reality underlying this Elysium was that the Department of Defense created a huge, integrated knowledge production enterprise aimed at achieving a particular desired outcome—victory over the Soviet Union. Similarly, the creation of stronger linkages between science and human well-being can be framed as an organizational challenge that requires a clear definition of the outcomes desired, and a mobilization of intellectual activity aimed at achieving these outcomes. If the resources and institutional structures are put in place, the science will happily follow.

Notes

1. I recognize that the precise meaning of the second and—especially— the third components have been and will remain infinitely contestable, yet this contesting is usually a matter of balance among agreed upon variables, rather than competition between mutually exclusive concepts.

2. For information on science budgets from the early Cold War years, see Office of Management and Budget, *The Budget of the United States Government, Fiscal Year 1999,* Historical Table 9.7 (Washington, D.C.: Government Printing Office, 1998); and National Science Foundation, *Federal Funds for Research and Development: Detailed Historical Tables, Fiscal Years 1951–1998,* at: www.nsf.gov/sbe/srs/nsf98328/start.htm.

There are many accounts of the political history of Cold War science and technology. Some of the better one's include: Leslie (1993), Zachary (1997), Norberg and O'Neill (1996), Lowen (1997), Sapolsky (1990), McDougall (1986), Kevles (1987), and Sherry (1977).

3. Ibid.

4. Biomedical and health-related research is the second largest component of the research portfolio, making up 22 percent of federal expenditures in FY 2000 (and 46 percent of nondefense research). For post-1960 research and development data, see National Science Board, *Science and Engineering Indicators* (Washington, D.C.: National Science Foundation), published biennially; and Intersociety Working Group, *Research and Development* (American Association for the Advancement of Science), published annually. Also see James Glanz, "Missile Defense Rides Again, *Science* 284 (April 16, 1999): 416–420.

5. For an introduction to some of these ideas, Jasanoff, Markle, Petersen, and Pinch (1995).

Bibliography

Agnew, B. (1999, March 26). NIH Invites activists into the inner sanctum. *Science, 283,* 1999–2001.

Ausubel, J. H. (1996, Summer). The liberation of the environment. *Daedalus, 125,* 1–17.

Bush, V. (1960). *Science, the endless frontier.* Washington, DC: Office of Scientific Research and Development. (Original work published 1945).

Cornwell, J. (Ed.). (1995). *Nature's imagination.* New York: Oxford University Press.

Glanz, J. (1999, April 16). Missile defense rides again. *Science, 284,* 416–420.

Graedel, T. E., & Allenby, B. R. (1995). *Industrial ecology.* New York: Prentice Hall.

Graham, L. R. (1998). *What have we learned about science and technology from the Russian experience?* Stanford, CA: Stanford University Press, 1998.

Holling, C. S. (1995). What barriers? What bridges? In L.H. Gunderson, C.S. Holling, & S.S. Light (Eds.), *Barriers and bridges to the renewal of ecosystems and institutions.* New York: Columbia University Press (pp. 13–14).

House Committee on Science. (1998, September 24). *Unlocking our future: Toward a new national science policy. A report to Congress.* Washington, DC: House Committee on Science.

Hudson, K. L., Rothenberg, K. H., Andrews, L. B., Ellis Kahn, M. J., & Collins, F. S. (1995, October 20). Genetic discrimination and health insurance: An urgent need for reform. *Science, 270,* 391–393.

Hull, D. L. (1988). *Science as a Process: An evolutionary account of the social and conceptual development of science.* Chicago: University of Chicago Press.

Jasanoff, S., Markle, G. E., Petersen, J., & Pinch, T. (Eds.). (1995). *Handbook of science and technology studies.* London: Sage Publications.

Kevles, D. J. (1987). *The physicists: The history of a scientific community in modern America.* Cambridge, MA: Harvard University Press.

Landy, M. K., Susman, M. M., & Knopman, D. S. (1999). *Civic environmentalism in action: A field guide to regional and local activities.* Washington, DC: Progressive Policy Institute.

Lane, N. (1996, February 9). Science and the American dream: Healthy or history. Speech presented at the Annual Meeting of the American Association for the Advancement of Science, Baltimore, MD. Available on-line: www.nsf.gov/od/lpa/forum/lane/slaaa.htm.

Lawler, A. (1998, June 12). Global change fights off a chill. *Science, 280,* 1683–1685.

Lee, K. N. (1993). *Compass and gyroscope: Integrating science and politics for the environment.* Washington, DC: Island Press.

Leslie, S. W. (1993). *The Cold War and American science.* New York: Columbia University Press.

Lowen, R. S. (1997). *Creating the Cold War university: The transformation of Stanford.* Berkeley, CA: University of California Press.

McDougall, W. A. (1986). *. . . . the heavens and the Earth: A political history of the space age.* New York: Basic Books.

National Science Foundation. (1998). *Federal funds for research and development: Detailed historical tables, fiscal years 1951–1998.* at: www.nsf.gov/sbe/srs/nsf98328/start.htm.

Norberg, A. L., & O'Neill, J. E. (1996). *Transforming computer technology: Information processing for the Pentagon, 1962–1986.* Baltimore, MD: Johns Hopkins University Press, 1996.

Office of Management and Budget. (1998). *The budget of the United States government, fiscal year 1999.* Washington, DC: Government Printing Office.

Pollack, R. (1997, August). Hard days on the endless frontier. *The FASEB Journal* 11, 725–731.

Sachs, A. (1996). Upholding human rights and environmental justice. In L. R. Brown & C. Flavin (Eds.), *State of the world 1996* (pp. 133–151). New York: W. W. Norton and Company.

Sapolsky, H. M. (1990). *Science and the Navy: The history of the Office of Naval Research.* Princeton, NJ: Princeton University Press.

Sclove, R. E. (1995). *Democracy and technology.* New York: Guilford Press.

Sclove, R. E., Scammell, M. L., & Holland, B. (1998). *Community-based research in the United States: An introductory reconnaissance, including twelve organizational case studies and comparison with the Dutch science shops and the mainstream American research system.* Amherst, MA: The Loka Institute.

Selznick, P. (1992). *The moral commonwealth: Social theory and the promise of community.* Berkeley, CA: University of California Press.

Sherry, M. S. (1977). *Preparing for the next war.* New Haven, CT: Yale University Press.

Subcommittee on Global Change Research. (1997). *Our changing planet: The FY 1998 U.S. global change research program.* Washington, DC: Office of Science and Technology Policy.

United Nations Development Programme. (1992). *Human development report 1992.* New York: Oxford University Press.

Warrick, J. (1998, October 28). AIDS's long shadow cools global population forecast. *The Washington Post,* p. A2.

World Health Organization. (1996). *World health report 1996.* Washington, DC: World Health Organization.

Wulf, W. A. (1998, September). Balancing the research portfolio. *Science, 281,* 1803.

Zachary, G. P. (1997). *Endless frontier: Vannevar Bush, engineer of the American century.* New York: The Free Press.

6

Is the "Citizen-Scientist" an Oxymoron?

STEPHEN H. SCHNEIDER

Is There a "Citizen-Scientist"?

Complex systems, like the earth's climate-ecosystem, will be characterized by a high degree of uncertainty and technical complexity for the foreseeable future. Climate is a good case study of the citizen-scientist question, and I will develop it here to make the arguments specific. Indeed, most everything that is interesting and controversial, like the behavior of complex systems that involve physical, biological, and social interactions, will never enjoy full understanding or full predictive capacity. Therefore these will always contain a high degree of subjectivity. Even for those systems that are objective, like a coin, people wanting to participate in policy making will have to learn to become comfortable dealing with probabilities. You cannot predict skillfully the sequence of faces of a multiply-flipped coin, but you can predict the odds of any sequence of faces—this is known as the "frequentist probability." The probabilities of each outcome are objective and you know what they are (at least for an unloaded coin). But estimating probabilities for interesting complex systems like climate and ecosystems or socioeconomic systems will involve a high degree of subjectivity—mixed in with elements of objectivity—this is often known as Bayesian or subjective probability. And even for the aspects of the climate problem for which we have lots of data and theory to make objective determinations, the way the data is applied often depends on assumptions which, in turn, are subjective.

Developing policy in an uncertain environment is a formidable task; a challenge made even more difficult by subjective assessments. What is the citizen, whether policy maker, journalist, or any nonspecialist in the lay public, to do in the face of this often bewildering complexity? To make effective policy, citizens need to ask scientists three fundamental questions. The first important question that a layperson can ask a scientist is What can happen?

103

The citizen tries to get experts—whether cancer specialists, military operations officers, environmental scientists, or economists—to agree on the range of possible outcomes. Honest experts will admit that surprises are possible—both happy and unhappy surprises—and that we have to anticipate those, too (but let's just leave our discussion here to the smaller universe of known outcomes). A typical outcome in the climate change debate could be three degrees warming if carbon dioxide in the atmosphere doubled.

But "what can happen" has little policy meaning by itself. What's more important is the *likelihood* of the event, so the second question is What are the odds? Consider three examples: (1) the probability of an asteroid hitting the earth, (2) the probability of correct identification using DNA fingerprinting, and (3) the probability of being killed for inner-city youth. The probability that the Earth will be hit by an asteroid that could wipe out 50 percent of the existing species and 99 percent of living things (other than bacteria), is exceedingly low, something like 1 in 10 million per year. This is comparable to the probability of being killed in a jet airplane crash. These odds are sufficiently low enough not to influence our actions very much. However, that is still a large number relative to some DNA fingerprinting introduced into evidence where 1 in 1 billion odds are claimed to represent a "reasonable doubt." In other words, citizens (jurors in this instance) have to learn to interpret what probability numbers mean. On the other hand, young men in the inner cities often face a probability of 1 in 100 of being killed. Yet too little is done, from a policy perspective, to address this situation. The point of these examples is to demonstrate that probability assessment is only one piece of information. The citizen must assign a value to these probabilities in order for the assessment to have meaning. The central, and critical, role of the citizen is to interpret these probabilities and determine an appropriate policy response. So, the citizen interested in science or thrust into a debate with scientific aspects has to learn to become comfortable with both objective and subjective probability conditions.

The final question is How do you know? When asking this question, the citizen is questioning the assumptions underlying debate. She is seeking an assessment of the key issues in the debate and an indication of the level of uncertainty surrounding any particular outcome.

The citizen-scientist is an important link between science and policy. The citizen-scientist is in a critical position to understand and use information from the scientific community to inform policy.

"Science" is not the Same as "Science for Policy"

In general, scientists strive in their scientific investigation to produce a large set of replicable experiments before probabilities (frequentist, in this case) can be attached to certain outcomes. Scientists are most comfortable when there is an empirical basis for those theories embodied in models used to project future outcomes like climate changes, population size, adaptive capacity, or endogenous growth of technology in response to climate policy. It is certainly true that "science" itself strives for objective empirical information to test theory and models. However, such objective or frequentist probabilities are not always available for policy makers to decide whether or how to respond to the implications of the state-of-the-art science (cf. Moss & Schneider, 1997).

An "objective" characterization of probability is the goal of most scientists, but often data is incomplete or other causes of uncertainty exist. Structural uncertainties, unobtainable data, or other impediments prevent the preferred situation in which all probability distributions can be "objective." Furthermore, even where there is a well-developed theoretical basis for believing certain outlier events of high consequence could occur (and perhaps even some empirical support for such a possibility), such outlier events often provide little basis for any objective assessment of probabilities. This causes many scientists to reject the notion of characterizing *any* likelihood estimates for such possible outcomes for which individual, corporate, local, and national decision makers have often chosen hedging strategies (e.g., purchase of personal insurance policies, corporate strategic investments, national vaccination programs or international security actions).

"Science for policy" must be recognized as a different enterprise than "science" itself, since science for policy involves being responsive to policy makers' needs for the best estimates of a wide range of plausible outcomes, even if those estimates involve a high degree of subjectivity. Most decision makers prefer to be informed about the wide range of possible events, and the levels of confidence the scientific community can assign to each event— as well as some estimates of how long it might take researchers to reduce those large uncertainties relative to how long it might take for such outcomes to actually unfold. For this reason, it is important to respond to the needs of policy makers and provide consistent and carefully labeled assessments of the wide range of outcomes.

Of course, uncertainty is not unique to the domain of climate change research. Even researchers in areas of science confined to the laboratory must confront uncertainties that arise from such factors as linguistic imprecision, statistical variation, measurement error, variability, approximation, subjective judgment, and disagreement. However, in climate research, as in other areas such as seismic hazard prediction, ozone depletion, and hazardous wastes, these problems are compounded by factors including their global scale, long lag times between cause and effect, low frequency variability with characteristic times greater than instrumental records and the impossibility of comprehensive experimental controls. Moreover, because climate change and other policy issues are not just scientific topics but also matters of public debate, it is important to recognize that even good data and thoughtful analysis may be insufficient to dispel some aspects of uncertainty associated with the different standards of evidence and degrees of risk aversion/acceptance that individuals participating in this debate may hold. Therefore, a "subjective" characterization of the probability will be the most appropriate. In this view, the probability of an event is the degree of belief that exists among leading experts that the event will occur given the information currently available.

Subjective Assessment: The "Climate Sensitivity" Example

Some experts do not like subjective assessment because they might be proved wrong. Consider the following scenario. Imagine a doctor who suspects that her patient may be suffering from a serious disease. She informs the patient of her subjective, preliminary opinion and suggests that diagnostic tests be performed to confirm or reject her opinion. These tests indicate a different diagnosis. It is well accepted that it would be dishonest and unethical for the physician to not inform the patient that the tests point to a new diagnosis. Yet, our political system seems to afford more credibility to people who predict the right answer (regardless of whether for correct reasoning) rather than those who got the answer wrong because of factors then unknown. How unknown factors turn out is luck, not skill. The "answer" is not really as important to a scientist as whether or not he or she gave the best judgment given what could have possibly been known at the time. Science does not assign credibility to people who got it "right for the wrong reasons"—the process is more important than the product. Science wants to know why we reach certain tentative conclusions. So should citizen-scientists.

Then, what is the reaction of a citizen (whether juror, judge, reporter, senator, or voter) when Scientist A says the probability of some catastrophic outcome is 25% and Scientist B says it is 2.5%. Citizens, even if statistically literate, can easily get confused, especially when each scientist uses long and complex technical arguments to back up their dissimilar intuitive, subjective judgments. This is typical of "dueling scientists." The resolution: *science-as-a-community* becomes important. Rarely can a few debaters be allowed to represent the credibility of a spectrum of views that characterizes the state-of-the-art knowledge base. A community of experts is needed to better accomplish that mission. Here is also where it gets tougher and tougher for the citizen to know how to participate. It is very difficult for an average citizen to listen to a technical debate among a few debating scientists, and really know whose subjective opinions about the likelihood of the various assumptions that

Table 1. Experts interviewed in the study

John Anderson, Harvard University	Michael MacCracken, U.S. Global Change Research Program
Robert Cess, State University of New York at Stony Brook	Ronald Prinn, Massachusetts Institute of Technology
Robert Dickson, University of Arizona	Stephen Schneider, Stanford University
Lawrence Gates, Lawrence Livermore National Laboratories	Peter Stone, Massachusetts Institute of Technology
William Holland, National Center for Atmospheric Research	Starley Thompson, National Center for Atmospheric Research
Thomas Karl, National Climatic Data Center	Warren Washington, National Center for Atmospheric Research
Richard Lindzen, Massachusetts Institute of Technology	Tom Wigley, University Center for Atmospheric Research/National Center for Atmospheric Research
Sykuro Manabe, Geophysical Fluid Dynamics Laboratory	Carl Wunsch, Massachusetts Institute of Technology

Expert numbers used in reporting results are randomized. They do not correspond with either alphabetical order or the order in which the interviews were conducted.

underlie disparate conclusions are more in the center of the knowledge spectrum than others. And even when there is a community assessment some will still claim that the process has neglected their contrarian evidence.

What does define a scientific consensus over the probabilities of possible events? Morgan and Keith (1995) and Nordhaus (1994) are two attempts by nonclimate scientists, who are interested in the policy implications of climate science, to tap the knowledgeable opinions of what they believe to be representative groups of scientists from physical, biological, and social sciences on two separate questions: first the climate science itself and second, impact assessment and policy. The first sample survey shows that although there is a wide divergence of opinion, nearly all scientists assign some probability of negligible outcomes and some probability of very highly serious outcomes, with one or two exceptions (e.g., scientist number 5 on Fig. 1 taken from Morgan and Keith (1995).

In the Morgan and Keith study, each of the 16 scientists listed in Table 1 was put through a several hour, formal decision-analytic elicitation of their subjective probability estimates for a number of factors. Figure 1 shows the elicitation results for the important climate sensitivity factor. Note that 15 out of 16 scientists surveyed (I am scientist 9) assigned something like a 10 percent subjective likelihood of small (less than 1°C) climatic change from doubling of CO_2. These scientists also typically assigned a 10 percent probability for

Schneider

Figure 1. Box plots of elicited probability distributions of climate sensitivity, the change in globally averaged surface temperature for a doubling of CO_2 ($2x[CO_2]$ forcing). Horizontal line denotes range from minimum (1%) to maximum (99%) assessed possible values. Vertical tick marks indicate locations of lower (5) and upper (95) percentiles. Box indicates interval spanned by 50% confidence interval. Solid dot is the mean and open dot is the median. The two columns of numbers on right-hand side of the figure report values of mean and standard deviation of the distributions. From Morgan and Keith (1995).

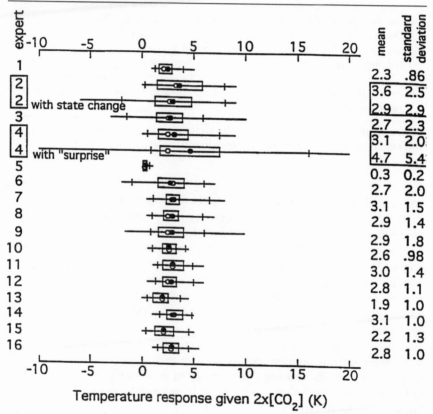

Temperature response given $2x[CO_2]$ (K)

extremely large climatic changes—greater than 5°C, roughly equivalent to the temperature difference experienced between a glacial and interglacial age, but occurring some hundred times more rapidly. In addition to the lower probabilities assigned to the mild and catastrophic outcomes, the bulk of the scientists interviewed (with the one exception) assigned the bulk of their subjective cumulative probability distributions in the center of the often cited range for climate sensitivity by mainstream assessment groups (e.g., IPCC, 1996). What is most striking about the exception, scientist 5, is the lack of variance in his estimates—suggesting a very high confidence level in this scientist's mind that he understands how all the complex interactions within the earth-system described above will work. None of the other scientists displayed that confidence, nor did the lead authors of Intergovernmental Panel on Climatic Change (IPCC, 1996). However, several scientists interviewed by Morgan and Keith expressed concern for "surprise" scenarios—for example, scientists 2 and 4 explicitly display this possibility on Figure 1, whereas several other scientists implicitly allow for both positive and negative surprises since they assigned a considerable amount of their cumulative subjective probabilities for climate sensitivity outside of the standard (i.e., IPCC, 1996) 1.5 to 4.5°C range for surface warming if CO_2 were to double. This concern for surprises is consistent with the concluding paragraph of the IPCC Working Group I, Summary for Policymakers.

IPCC lead authors, who wrote the Working Group I Second Assessment Report, were fully aware of both the wide range of possible outcomes and the broad distributions of attendant subjective probabilities. After a number of sentences highlighting such uncertainties, the report concluded: "Nevertheless, the balance of evidence suggests that there is a discernible human influence on the climate." The reasons for this now-famous subjective judgment were many. These include a well-validated theoretical case for the greenhouse effect, validation tests of both model parameterizations and performance against present and paleoclimatic data, and the growing "fingerprint" evidence that connects observations of horizontal and vertical patterns of climate change to the patterns predicted to occur in coupled atmosphere-ocean models. Clearly, more research is needed, but enough is already known to warrant assessments of the possible impacts of such projected climatic changes and the relative merits of alternative actions to mitigate emissions, make adaptations less costly, or both. That is the ongoing task of integrated assessment analysts (e.g., Schneider, 1997b), a task that will become increasingly critical in the next century. To accomplish this task, it is important to recognize what is well established in climate theory and modeling and to separate this from aspects that are more speculative. That is precisely what IPCC (1996) has attempted to accomplish. What if we are left with dueling experts without the benefit of a fair representation of the spectrum of beliefs? For example, in the climate change debate, what if there were no IPCC report? The job of the citizen-scientist is facilitated when institutions like IPCC exist that promote "science as community." In the absence of such community assessment, the job of the citizen-scientist becomes more difficult.

Copernican Revolutions are Rare

Past episodes of basic changes in scientific thinking—*paradigm shifts*—are frequently invoked by contrarian scientists and supporting politicians to argue that the consensus view might turn out to be false. Indeed, many famous examples in the history of science demonstrate that new discoveries, or new theories reinterpreting well-known evidence, can displace apparently well-established knowledge. The two most cited examples are the overthrow of geocentrism or how Einsteinian relativity displaced Newtonian mechanics. Despite the fact that most basic theories (e.g., laws on conservation of mass, momentum, and energy) have stabilized in the latter half of this century, such paradigm shifts can still be expected to continue. History provides no reason to suppose that many more seemingly radical theories will not occur, even if we are forced to endure more "end of" popular books sporting polemical titles like the *End of History* or *The End of Science*.

At the same time, within the history of any given discipline the number of real paradigm shifts—truly fundamental changes in concepts, methods, and conclusions, sometimes called "scientific revolutions"—has generally been quite small. Thomas Kuhn (1962), originator of the concept, observed that "normal science," which consists of extending and analyzing the dominant paradigm's observations, experiments, and models, makes up the vast bulk of scientific activity. Just as in advertising, the all-too-frequent claims of "revolutionary" change are usually no more than rhetoric. Rarely are "Bargain Antiques" much cheaper or "Painless Dentistry" less uncomfortable than the mainstream competitors. Furthermore, the vast majority of paradigm challenges either fail or are really minor adjustments to normal science. (One recent case in point is the ill-fated claims of "cold fusion.")

It is true, however, that brilliant, revolutionary ideas have sometimes been ignored merely because they did not fit existing modes of thought, and it is virtually certain that this is happening now, somewhere, and will happen again in the future. Yet it is also true that most paradigm-challenging ideas fail because they are—to put it bluntly—simply wrong. Just as with battles, history tends to lionize successful scientific revolutionaries while fast forgetting the far larger number of failures.

Let me rephrase this in two points. First, being in the center of the current knowledge spectrum does not mean that you are right. Remember, Ptolemaic supporters with long beards and flowing robes, holding high positions in church and state, held sway for many centuries with the wrong theory and dissenters faced great personal peril—as the Galileo story exemplifies. However, most problems are not Copernican geocentrism. For most problems, the scientific mainstream is not a paradigm away from truth. And even though a few problems will turn out to be Copernican, for every real

Copernicus there are probably a thousand hopeful contenders. Only rarely does someone with great convictions and correct insights stand up to the scientific establishment and overthrow the dominant paradigm. This does not mean that alternative views or paradigms should be ignored, but should be treated with the same skepticism afforded all work (Schneider, Turner, & Morehouse-Garriga, 1998).

But how can the citizen figure out where the mainstream is and whose subjective probabilities to trust? The "citizen-scientist" becomes an oxymoron when facing this dilemma unless he or she is willing to put in a lot of effort: going to the library, reading National Research Council studies, Intergovernmental Panel on Climate Change assessments, and so on—in essence, trying to understand whether the subjective opinions of dueling experts represent marginal or mainstream views. And figuring out whether the marginal views are truly breakthrough (or in fact anything other than ideological blindness or special interest disinformation) is a tough nut to crack. I was once accused in debate of advocating "science by consensus rather than by experiment." After all, no Galileos would have emerged if only the consensus opinions were trusted, I was reminded. "But for decades I have believed in science by experiment" (e.g., Schneider & Mesirow 1976, p.10), I replied. All things in science are forever questioned and questionable. However, without inconsistency, I continued, "I also believe in science policy by consensus." Society is prudent to go with the best guesses of the majority, always keeping in mind the following three questions—and always challenging the dominant paradigm—while acting as if it were still likely to be true. This broader role for science in the context of social issues has been separated from the "normal science" of Kuhn and relabeled "postnormal" science by sociologists of scientific knowledge (cf. Funtowicz & Ravetz, 1992; Funtowicz & Ravitz, 1985; Jasanoff & Wynne, 1998).

The Three Questions for the Environmentally Literate

To recap, question 1 is What could happen? and question 2 is What are the odds? Question 3, then, is How do you know? In the process of a citizen asking an expert How do you know, one is asking to get deeply involved in the details of the debate. For example, I would be skeptical if a scientist presented me with a "certain truth" which ignored or dismissed underlying assumptions and their uncertainties. Cock-sureness about complex issues undermines credibility and is a telltale sign that it is time for a second opinion. But even that determination can be difficult for the citizen who is not very familiar with typical scientific debates. Sophisticated ability to discern who is more credible requires a citizen that is more than casually interested, but who is passionately involved. It requires one who watches Nova, or religiously reads *Scientific American* or the Tuesday Science section of the *New York Times* or researches issues in the library to ferret out credible views from marginal claims. Rarely is this distinction discernible in the popular media or by engaging a typical search engine on the internet with a few key words.

True scientist-citizens include good science journalists or the few senators and ministers who attend the bulk of their own hearings when scientists appear in detailed debates. I think U.S. Vice-president Gore is in that category, as is the former U.S. Senator from Colorado, Tim Wirth, or former prime minister of Great Britain, Margaret Thatcher. Such people have learned enough of the process of science over the years to become like a good science reporter: they have a nose for phonies or for people who are passionate without balance, and I do not mean "balance" in the traditional journalistic sense of pitting one extreme versus the mainstream (or sometimes versus the opposite extreme) without labeling where each claimant sits in the spectrum of knowledgeable opinions. Such inappropriate balance usually leaves nearly everybody not deeply familiar with the debate and the debaters confused.

A true citizen-scientist meets some pretty tough criteria that would exclude all but a small, dedicated minority. The group of people who can competently answer the three fundamental questions is limited. Ideally, a larger, broader group of people would have access to answers to the three questions.

Meta-Institution Building: A Science Assessment "Court"?

Being able to judge a scientist's credibility is a tall order for most members of Parliaments or Congress—or even their technical staffers. So what I think we need, at the risk of sounding somewhat elitist, is a "meta-institution." We need an institution in between laypeople and the expert community to help the citizen sort out conflicting claims. I am not implying that such a meta-institution is there to help citizens decide, for example, whether or not nuclear power is safe or whether we should have a carbon tax. Those are personal value judgments based on each person's political philosophy as to whether the risks and costs of one energy system justify turning to another, or whether investing current resources is worth the expenditure to reduce risks of performing the climate change experiment on "Laboratory Earth" (Schneider, 1997a). I believe that these are value judgments every citizen is already equipped to make—if only he or she knew what can happen and what the odds are.

To repeat, citizens need expert help with the "what can happen" and "what the odds are" parts of policy making. That is, laypeople need guidance in figuring out where the mainstream expertise lies and how confident it is of important outcomes—and that is where we may need some new institutions.

We actually have had institutions with the goal of performing such assessments of complex issues. One such was recently killed by the U.S. Congress: the Office of Technology Assessment. It was a nonpartisan congressional assessment office whose job it was to provide reports that could cut through the claims and counterclaims of the special interests, faxed by the thousands to the halls of power and the media—what I label the "one fax one vote syndrome." Too often, lobbying groups in the name of free speech buy bigger and bigger megaphones for nonmainstream experts whose "science" ostensibly supports their positions. If they get heard enough, these amplified experts might get equal status at the bargaining table with those espousing consensus views—that is, repeated exposure often buys equal credibility for what really is not a very credible scientific position by simply overwhelming the communications apparatus and scientific literacy of most citizens with complex technical counterexamples.

Often, such disinformation specialists are counting on citizens not being able to sort out the credibility of conflicting technical claims for themselves. This is common; every citizen is used to it. Nevertheless, it is very baffling. Staffers at the Office of Technology Assessment were not fooled by this strategy, nor have been National Research Council committees or Intergovernmental Panel of Climate Change experts. What these assessment groups do is evaluate the credibility of various conflicting claims. They do not decide policy. Policy determination involves personal value judgment about the response to a set of possible consequences, including assessment of priorities, risk, and costs. The expert assessment teams decide *whether the claims made about those probabilities are credible* or at least what the subjective probability estimates might look like—as the Nordhaus (1994) experts did. This task cannot be done easily by lay citizens—including journalists and decision makers.

That is where I separate out the role of the citizens from the experts. Typically, the experts assess the odds, but citizens make the choices about how to take risks of various kinds. If we set up a meta-institution, something in between the citizen and the scientist, it is absolutely essential that it have one characteristic—openness and transparency to all citizen groups, including special interests. For credibility, the assessors cannot meet behind closed doors where they would be in greater danger of scientific elitism or personal value biases creeping into their ostensibly balanced scientific judgments. Unfortunately, there are scientists who believe that it is inappropriate for scientists even to discuss in public issues in which there is a high degree of uncertainty. I think that is the truly elitist position, because what it says is that we scientists should be the judges of *when* we tell the public what is possible and at what odds. I think that openness is essential: reporters need to be there, special interests need to be there, ordinary citizens need to be there as *witnesses to the assessment process itself* (Edwards and Schneider 2000). Citizens' roles are not to determine what the probabilities of various claims are because that's not their competency, but to make sure that the assessment process is open and to ask the right questions.

A "FED for Science"?

The most difficult aspect is how we choose who sits on this "science assessment court." This is not a court that assesses guilt and innocence or makes policy recommendations, but a court (see also Kantrowitz, 1967) in a sense that it can evaluate the probability of claims. I propose the assessment team's members should primarily be chosen from the lists of the scientific societies (National Academy of Science, American Physical Society, Ecological Economics, etc.), perhaps with a few spots reserved for political appointees. The tenure of such appointees should survive the electoral terms of presidents and senators: perhaps a ten-year term. In a sense, I am proposing an institution like the Federal Reserve Board. However, my "FED for Science," would be *only* an information agency, unlike the FED, which actually makes economic policy. Its job is to label the exaggerators, the distorters, and the passionate who just cannot (or will not) see past their own denial of special interest. Perhaps the best metaphor is Consumers Union, a private watchdog and ratings service for products and services that independently assesses claims by producers or providers against objective and subjective tests.

As I suggested, citizens and interest groups should definitely be principal agenda setters and witnesses to the new institution, but not the assessors. This is my concrete proposal for getting the public more involved in debating technical issues. If the public is largely confused by a baffling technical brouhaha, then instead of participating in the value-laden, policy choice process, they are more likely to abdicate everything to the experts. That withdrawal of citizens from the decision-making process leaves the field wide open to the special interests to compete over who can shout louder, take out bigger ads, whose fax machines have a higher baud rate, or who can best finance which congressional representative's campaign—all in order to put the "spin" on scientific knowledge most favorable to their own interests. I would like to move the process of evaluating scientific credibility away from the political arena and into a meta-institution that has no responsibility for policy choice, has no decision-making authority, but can call anyone's statement, including one from the President or the Speaker of the House, "scientific nonsense."

It is an act of courage for an expert employed in the Executive branch to come out and say, "I'm sorry, the facts of my President are wrong." Such acts of courage are, almost by definition, discouraged by the hierarchical nature of the political chain of command; that is why we need an independent information agency. Information cannot be under the exclusive control of people with vested interests in the answer. All interests should be witnesses to the process, they should be able to ask questions of the assessors, should be able to have their favorite "new Copernicus" able to testify to the assessors, but they should not be able to vote on the credibility of the conclusions.

As I stated earlier, we already do have institutions that do scientific assessments. One problem for the National Research Council (NRC) is that it is under political attack from people who claim it is "elitist" because there are no citizens in the process—other than to fund their studies. A second problem is the NRC only operates when somebody gives them enough money to do an in depth examination of very specific questions. Moreover, the sessions are typically closed, with agendas set by the study members and the funding agency. The meta-institution I envision here is a new body that can take a letter from an ordinary citizen baffled by some conflicting Op-Ed essays, from a congressperson skeptical of a scientific witness, from an environmental group or an industrial lobby and provide a considered (say 6 weeks) reaction, not a two-year focused study. This "science assessment court" would partly rely on existing NRC studies for input, but not try to duplicate them. The assessors would basically be in the position of evaluating quickly the credibility of a whole series of claims and counterclaims about the validity of some scientific proposition or the probability that some outcome will occur. Moreover, they could commission an NRC study, or at least they could ask Congress to do that.

I do not envision a massive new bureaucracy, although some permanent staff are obviously needed. The design might stress a network of experts not all sitting in a formal edifice, but who are willing on short notice, to spend several days each week for a month to write a report on the credibility of specific issues, and their meetings would be available to be witnessed either in person or perhaps by closed circuit TV. I do not claim to know what the right institutional architecture is, nor the degree to which it should be internationalized to deal with global environment-development sustainability issues. I propose the concept of a meta-institution primarily because the current cacophony of claim and counterclaim contributes to the disenfranchisement of citizens from the scientific process. If my specific proposal is considered unworkable by some, then they should please propose a better one—the status quo simply is not working well.

In my first book, *The Genesis Strategy* (Schneider & Mesirow, 1976), I proposed a "Truth and Consequences Branch," a fourth branch of U.S. government, where people would be appointed for 20 years in staggered terms. Then, I envisioned a much higher visibility bureaucracy, whose job was to expose the phony scientific claims of the government. Now, I am more concerned about policy that is too often proposed on the basis of junk science (e.g., Ehrlich & Ehrlich, 1996).

Undoubtedly, some will attack this idea on the grounds that a new institution is an unnecessary federal expenditure to create yet another agency which will engage in inappropriate interference with the constitutional privileges of citizens to advertise their perceived brand of "truth" as loudly and effectively as they can. After all, such defenders of infomercials will claim, "both sides" are free to promote their elliptical views. I do not propose banning their faxes or their advertisements. However, this does not obviate the need for a balanced partisan-free presentation of the issues surrounding the debate. I do not think there is anything wrong with using public financing to evaluate the credibility of the scientific components of their various claims and counterclaims.

Advocacy Yes, Selective Inattention to Facts, No

I personally dislike the way the courtrooms often use expert witnesses. It is a very bad way to get toward the truth when each side picks elliptical experts who do not believe it is their job to make their opponent's case. Moreover, I think that is a scientifically unethical epistemology. I believe it is the job of an honest scientist to examine *all* plausible cases, and then to provide a subjective probability for each conceivable outcome that honestly reflects the range of information each expert believes to be most credible. Now, an expert could have a personal opinion on what to do with this probability assessment, which raises the question: "Is the scientist-advocate an oxymoron?" I believe this dual role is not an oxymoron, but requires great care. The scientist-advocate must work hard to separate out the factual from the value components of a debate. But an unconscious prejudice can be worse than a conscious one because if your prejudice is unconscious you cannot even fight to fix it. At least conscious prejudice creates the opportunity to change. With unconscious prejudice or ideological zeal, advocates can be captured by their perceived brand of "truth."

To me, the best safeguard for public participation in science-based policy issues is not to leave subjective probability assessment to a few charismatic individuals, but to the larger scientific community. Some will say it is impossible for an expert to be in a public debate with policy overtones and retain his or her objectivity in the science. I think that is not necessarily true. Just because some people cheat does not mean all do. No one is exempt from prejudices and values. The people who know it and make their biases explicit are more likely to separate them from their sober judgments about probabilities than those professing to be value neutral. If you *knowingly* distort what you believe to be the likelihood of certain outcomes for ends-justify-the-means reasoning, you are no scientist, you are just dishonest.

However, in the real world no one—expert or layperson—gets all the time needed to explain every nuance of complex issues. We are forced to be selective or be ignored. I have called this the "double ethical bind" (e.g., Schneider, 1989). To deal with the dilemma of trying to be heard but not to exaggerate, I focus on aspects of a debate which convey both the urgency and the uncertainty, typically by using metaphors. For example, in the climate change debate, I could use a metaphor of a low probability of getting cancer. But I think that metaphor exaggerates the risk. If the worst happens with cancer, you die, and I do not see global warming as killing all of us or Nature. Rather, I see it as a potentially serious stress that threatens selectively. A much better metaphor therefore, would be a parasite, or a debilitating condition. But this metaphor may seem inappropriate to some because it is not an exact match to climate change risks. So we are always stuck on this treacherous ethical ground between finding metaphors that simplify the complexity of the problem, yet accurately convey the risks and uncertainties of the case. If scientists do not find the metaphors to communicate, most citizens simply will not hear them. Instead, they will hear the infomercials, ads, and press releases faxed to journalists everywhere by those who do not think it is their job to make their opponent's case.

When an expert communicates with metaphors and is willing to play in the sound-bite world, even though, like me, you might be uncomfortable, there is one more step you can take to be as responsible as possible. Public scientists or scientific bodies that make public statements should also produce a hierarchy of backup products ranging from Op-Ed pieces (which are often a string of written sound bites), to *Scientific American*-length popular articles that provide more moderate depth, to full-length books coming out every five years or so. Such books should document in great detail the aspects of an issue that are well understood and separate these from those that are more speculative. Books also should provide a detailed account of how one's views have changed over time as the scientific evidence has changed. And even if only a decreasingly small segment of the public really wants to know what you think in detail about the whole range of questions, at least you retain an ethical stance because you have made available to anyone interested via articles and books in the popular and scientific literatures the closest one can come to full disclosure of what's known and uncertain—which is required of honest science. But full disclosure is simply not possible in time-constrained congressional or media debates—metaphors have to do the job, and the hierarchy of backups are crucial for full disclosure.

Rolling Reassessment

What happens when the current state of science has missed something really big that is potentially dangerous; or something we currently fear proves unfounded. That is why another element to this process must always be added: what I call "rolling reassessment" (Schneider, 1997a). It takes immediate actions to reverse long-term risks, but such actions are not without costs. Therefore, we should initiate flexible management schemes to deal with large-scale, potentially irreversible damages, and allow ourselves to revisit the issue every, say, five years. Credibility is not static—there are new outcomes to be discovered or other ones we can eventually rule out. With such new knowledge, whose credibility could be partially reassessed by the meta-institutions I proposed earlier, the political processes could decide we did not move fast enough in the first place (or maybe we moved too fast) and consequently make adjustments. The trouble is that once we have set up certain kinds of fixed political establishments to carry out policy, people can become vested and reluctant to make adjustments, either to the policies or the institutions.

Environmental Literacy

The long-term solutions to making competent citizen-scientists involves more than a meta-institution to evaluate the credibility of conflicting claims. We need competent consumers of the assessments generated by the scientific assessment meta-institutes. It involves education. It involves environmental and scientific literacy.

We rarely teach science literacy in school, even when we think we do via "science distribution requirements." Literacy is not just knowing the *content* of some scientific disciplines, as important as that is. It is not practical to teach detailed scientific content of a dozen relevant disciplines to all citizens—and it is not even necessary. What citizens need to know is the difference between a factual and a value statement; the difference between objective and subjective probabilities; the difference between a paradigm and a validated theory; the difference between a law and a system; the difference between a phenomenological model and a regression model (by that I mean the difference between associations of two data sets and a validated theory). Many people think that a correlation between two variables indicates causation or predictive power. A correlation is not a law. Just because the association predicted correctly a few times, does not mean it will always do so. Credible predictions come from having modeled the causal process correctly, not just from extrapolation of a few correlations. When the future conditions are different than the conditions in which some correlation was first observed, a process model will likely out perform a strict empirical model. Finally, environmental literacy means knowing the social process of knowledge transfer (i.e., media) and the political process through which decisions are made (cf. Schneider, 1997c), including an ability to sort out which claims and counterclaims by the one-fax-one-vote folks are more credible—that is where the meta-institutions like a "FED-for-science" come in.

Citizens need to have a high level of environmental and scientific literacy, but we rarely teach it in formal education. I would like to see elementary schools teaching these concepts, teaching by examples and via dialogues with students, how to separate facts and values, the difference between objective and subjective probability, efficiency versus equity considerations and conservation of nature versus economic development trade-offs. It could be done. I think what environmental literacy can do is empower citizens to begin to pick a scientific signal out of the political noise that all too often paralyzes the policy process.

Bibliography

Edwards, P. N., & Schneider, S. H. (forthcoming 2000). Self-governance and peer review in science-for-policy: The case of the IPCC second assessment report. In C. Miller and P. N. Edwards (Eds.), *Changing the atmosphere: Expert knowledge and global environmental governance.* Cambridge, MA: MIT Press.

Ehrlich, P. R., & Ehrlich, A. H. (1996). *Betrayal of science and reason: How anti-environmental rhetoric threatens our future.* Washington, DC: Island Press.

Funtowicz, S., & Ravetz, J. (1985). Three types of risk assessment: A methodological analysis. In C. Whipple & V. Covello (Eds.), *Risk analysis and the private sector.* New York: Plenum.

Funtowicz, S., & Ravetz, J. (1992). Three types of risk assessment and the emergence of post normal science. In S. Krimsky and D. Golding (Eds.), *Social theories of risk.* London: Praeger.

Intergovernmental Panel on Climatic Change (IPCC). (1996). Houghton, J. T., Meira Filho, L. G., Callander, B. A., Harris, N., Kattenberg, A., & Maskell, K., (Eds.). *Climate change 1995. The science of climate change: Contribution of working group I to the second assessment report of the Intergovernmental Panel on Climate Change.* Cambridge: Cambridge University Press.

Jasanoff, S., & Wynne, B. (1998). Science and decisionmaking. In S. Rayner & E. L. Malone (Eds.), *Human choice and climate change,* vol. 1. Ohio: Batelle Press.

Kantrowitz, A. (1967, May 12). Proposal for an institution for scientific judgement; excerpts from a report to U.S. Senate, March 16, 1967. *Science, 156,* 763–764.

Kuhn, T. S. (1962). *The structure of scientific revolutions.* University of Chicago Press, Chicago.

Morgan, M. G., & Keith, D.W. (1995). Subjective judgments by climate experts. *Environmental Science and Technology,* 29, 468A–476A.

Moss, R., & Schneider, S. H. (1997). Characterizing and communicating scientific uncertainty: Building on the IPCC second assessment. In S. J. Hassol & J. Katzenberger (Eds.), *Elements of change.* Aspen, CO: AGCI.

Nordhaus, W. D. (1994). Expert opinion on climatic change. *American Scientist, 82,* 45–52.

Schneider, S. H. (1989). *Global warming: Are we entering the greenhouse century?* New York: Vintage Books.

Schneider, S. H. (1997a). *Laboratory Earth: The planetary gamble we can't afford to lose.* New York: Basic Books.

Schneider, S. H. (1997b). Integrated assessment modeling of global climate change: Transparent rational tool for policy making or opaque screen hiding value-laden assumptions? *Environmental Modeling and Assessment, 2,* 229–249.

Schneider, S. H. (1997c.) Defining and teaching environmental literacy. *Trends in Ecology and Evolution, 12* (11): 457.

Schneider, S. H., & Mesirow, L. E. (1976). *The genesis strategy: Climate and global survival.* New York: Plenum.

Schneider, S. H., Turner II, B. L., & Morehouse Garriga, H. (1998). Imaginable surprise in global change science. *Journal of Risk Research, 1*(2), 165–185.

7

Should Philosophies of Science Encode Democratic Ideals?

SANDRA HARDING

External versus Internal Democracy Issues

How are the economic benefits and costs of the production of scientific information distributed within societies and between them? Who receives the social and political benefits and costs? Who gets to make the decisions that produce such distributions? Are the processes responsible for such distributions democratic?

Most people concerned to strengthen the links between modern science and democratic projects have focused on these kinds of questions, raising issues external to sciences' cognitive, technical core. According to this external democracy view, as it will be called, sciences' internal core—its most tested theories, models, methods, descriptions, and explanations of nature's order—is immune to such questions. These people have been concerned, for example, with who has access to mathematical, technological, and science training, who gets to decide which scientific projects should be funded, who gets access to the information and technologies that research makes possible, and who gets to make decisions about the social and environmental risks generated by scientific and technological projects.

Such external issues arise in global as well as national contexts. They have emerged in controversies over the role of sciences (intended or not by scientists) in military matters, environmental destruction, and the negative consequences of development policies and practices on the world's economically and politically least-advantaged groups. For example, U.S. funding for the natural sciences—physics especially—disproportionately has been tied to national security priorities. Yet the military information and technologies produced end up disproportionately used within (or against) third world societies in Latin America, the Middle East, and Africa. In another case, post–World War II development projects were supposed to enable third world societies to reach the higher standards of living available in the first world. Development was to be accomplished through the transfer to the third world of first world sciences, technologies, and their philosophies of rational inquiry and rational organization. Yet the priorities, policies, and practices of these programs have ended up "developing" primarily the already most economically and politically well-positioned groups in the North and in third world countries. They have largely dedeveloped and maldeveloped the great majority of the worlds' peoples who are already the most economically and politically vulnerable. In most cases development policies and practices redirected the third world's natural resources and human labor to serve the needs of trans-national corporations and "the investing classes" in the first and third worlds, and substituted socially and environmentally destructive lifestyles for less harmful local practices (Harcourt, 1994; Sachs, 1992; Sparr, 1994). Modern sciences' agendas often have ended up aligned with antidemocratic projects globally as well as nationally, though this certainly was not the intent of most of the scientists or, in many—perhaps most—cases, of the development administrators. For the most part, preserving and/or advancing desirable cultural, political, and environmental values were simply not on the agenda of development agencies or their funders.

Efforts to promote external democracy, crucially important as they are, do not challenge the idea that social and political neutrality can, does, and should characterize sciences' internal, cognitive, technical cores. They do not challenge the Enlightenment assumption that sciences can be, should be, and in the best existing cases are value-free. Of course modern sciences are conducted within social worlds in that their human and material resources must be provided by the larger social order. Moreover, the amounts, kinds, and sources of such resources have varied from era to era and from one culture to another. But sciences' trans-cultural, socially neutral theories, models, and methods are believed to enable them to detect the facts about the order of the universe that are everywhere and always the same. According to the externalists, sciences are in society, but society is not *in* sciences, their best theories, models, methods or results of research.

In contrast, cognitive democracy approaches, as they will be called, are concerned with how social and political fears and desires get encoded in that purportedly purely technical, cognitive core of scientific projects. How best to deal with this kind of phenomenon in trying to link sciences more closely to democratic projects is the topic of my analysis here. However, before turning to say more about it in the next section, several possible misunderstandings of this kind of project must be addressed.

Some readers may fear that this cognitive approach adopts a relativist epistemology, or consists in a "flight from reason." There are no sound reasons for such fears. The cognitivists' point—at least as I and the vast majority of such analysts develop it—is not that scientific practices are "nothing but" social, political projects, or that the representations of nature they produce are shaped entirely and only by such projects. Rather the point is that technical, cognitive elements of scientific practices and the information these produce always represent social and political priorities, meanings, and ideals as well as more or less accurate pictures of nature's order. (Central arguments in this approach will be reviewed below.) There are indeed rational theoretical and practical standards for evaluating competing knowledge claims. Modern sciences do in many obvious ways achieve less and less false claims about nature's order.

It is easy to overlook the fact that "less and less false" is not the same as "true." One can never be sure the sciences have arrived at absolutely true claims for two reasons: present claims must be held open to revision in case of the appearance of further empirical evidence, and they must be held open to the need for fruitful conceptual shifts. It is these considerations—these reasons for refusing to claim the production of Truth—that are supposed to distinguish empirical sciences from dogmatic positions. Most of the observations made by medieval astronomers are still facts within astronomy today. Yet the hypothesized relations between these facts and the meanings such facts have in the modern world are vastly different from the relations and meanings attributed to them in the medieval world.

Philosophies of science are one site where social and political fears and desires appear. Such philosophies are produced, used, interrogated, revised, and refined by professional philosophers, but also always by scientists, science policy-makers and managers, and by all of us citizens who consume the products of science in the air we breathe, the food we eat, the technologies of everyday life, and myriad other ways. That is, philosophizing on any topic rarely is monopolized by professionals since other groups usually have their own interests in the kinds of general and prescriptive theories about human activity on which professional philosophers professionally focus.

Concern with philosophies of science may seem to many readers largely irrelevant to the real life projects of encouraging closer links between modern sciences and democratic policies and practices. My point is that it is precisely such encoded ideals that have powerful effects on our daily activities. Such ideals make seem natural, logical, "common sense," and otherwise desirable precisely the kinds of antidemocratic policies and practices of concern to the externalists. There are consistencies between the external antidemocratic policies and practices that shape modern sciences and models or idealizations of such policies and practices that can be found in philosophic aspects of sciences' cognitive cores. Moreover, this argument makes another point: that there are significant scientific costs to such antidemocratic idealizations as well as the more obviously recognizable political costs.

Another caveat. Perhaps such concerns will appear to *introduce* social and political elements into otherwise socially neutral sciences and accounts of them. As will be clear from what follows, however, everyone who reflects on the matter understands that modern sciences already do encode precisely such social and political fears and desires. The question here, instead, is whether they should more effectively encode democratic ideals, how they can do so, and on what grounds could one justify such recommendations. A preoccupation with the futile (and undesirable, the argument will go) project of excluding social and political fears and desires from the sciences' cognitive cores delegitimates and distracts from these kinds of important issues.

There are many respects in which philosophies of science could encode democratic ideals. Here I shall focus on just one. Elsewhere I have argued that the universality ideal in particular is scientifically and politically dysfunctional (Harding, 1998, chapter 10). Pursuit of the universality ideal devalues cognitive diversity, which is now and always has been an important resource for the growth of knowledge. The sections that follow will summarize those arguments, and then propose that something important for democratic sciences can be retained and transformed from the older universality ideal nevertheless. We can retain the recognition that it is valuable to try to universalize one's hypotheses—to see how far from their cognitive and cultural point of origin they have value. However, the desirability of such processes should themselves be universalized. An appropriate philosophy of science should encourage the recognition that many scientific and technological traditions besides those of modern science contain elements that can prove valuable far from their site of origin. Moreover, it is important that no one culture's scientific and technological traditions serve as a gatekeeper on the flow of such processes and the cognitive resources they distribute. Thus, a philosophic model that recommends appreciation of the distinctive strengths of many cultures' knowledge systems for the legacy of human knowledge can link cognitive and political virtues in a more extensive encoding of the democratic ethos within philosophies of science.

Encoded Democratic Ideals

The practice of trying to identify cultural, social, and political elements in scientific projects brings them to our attention and makes possible critical examination of them. Sciences have an "integrity" with their historic eras, as Thomas Kuhn famously put the point, and as the subsequent post-Kuhnian, feminist, and postcolonial science and technology studies have demonstrated

(Harding, 1998). Such practices make these elements objects of our conscious thought, thus preventing them from functioning as unexamined evidence for the apparent reasonableness of one claim over another. We can learn how philosophies of science that guide research already do encode political fears and ideals.

However, not all of such fears and ideals are ones we should want to eliminate. Some are democratic, and these democratic elements are thought to advance the growth of knowledge. For example, everyone regards it as a strength, not a limitation, that philosophies of science insist that the social status of the observer should not provide a standard for evaluating the adequacy of scientific observations or arguments. Modern science is supposed to be a democracy in this sense, not an aristocracy. Galileo argued that anyone could see through his telescope the facts about the heavens. So, too, scientific practices today support the assumption that a graduate student no less than a Nobel Prize winner can turn out to be the one who comes up with an important observation or argument. Of course experienced observers can be expected to come up with more reliable contributions than those less experienced. The constant emergence of valuable observations and arguments from less experienced observers, however, sustains belief in the scientific value of this democratic ideal.

Relatedly, the results of research must be replicable by any individual or group that wishes to undertake the rigorous procedures necessary to produce them. This requires that the results of research must be public; they belong to "humanity," and on scientific grounds may not legitimately be shielded from public view. There is a related external-democracy issue with which this one may be confused. At times it has explicitly been claimed that the results of research should also be available for anyone to use; scientific knowledge should be public in this sense also. Appeal to such democratic principles can still be detected in arguments for funding projects promised to advance "human knowledge," to increase "human welfare," or to contribute to "human progress." And groups who receive too few of the benefits and who bear too many of the costs of scientific research often make this claim. Yet systems of contracts, patents, and licenses now insure that the results of scientific research that have the greatest social consequences are not public in this sense. They are privatized by those groups powerful enough to enforce such monopolies, such as states, corporations, and the research institutions that they sponsor. Thus, in many respects citizens who are not privileged to be party to such contracts, patents, and licenses now have the least access to the results of research that have the greatest consequences for their lives. This is an issue about the social "uses and abuses" of scientific information, however. My point here is that the required publicity of the results of research is an important democratic ideal that is part of scientific method; it is part of the cognitive, technical core of scientific practice.

Again, in early modern science the belief that matter was everywhere composed of the same kinds of materials was perceived to be a radically democratic claim. It challenged the Christian argument that the celestial and terrestrial spheres were composed of fundamentally different kinds of matter that obeyed different kinds of laws. To challenge the hierarchy of matter was to challenge also the political hierarchies that the Christian view of matter modeled. Christian hierarchies began to lose their ability to model the "order of nature," and vice versa, after the advent of early modern science, for the matter composing the earth and the heavens increasingly was seen as identical and thus equal. This case, too, demonstrates that there is nothing new about encoding political ideals in the cognitive core of the sciences— this time in a scientific ontology. This is not an example of political neutrality winning out over a politics of science, but rather of one politics replacing another. The "equality of material bodies" is or should be a political principle of democracies, as feminist and race theorists have been arguing. What is at issue here instead is whether the political ideals currently encoded in the cognitive, technical cores of the sciences are the ones we do and should want.

My point is not that a "politically correct" philosophy of science automatically insures better sciences, a more democratic social order, or a more effective link between them. Rather, the social ideals encoded in scientific thought about nature and in technical and popular thought about the sciences provide justifications for external, political practices, including desirable political practices. The values and interests that appear in sciences' cognitive cores can have positive scientific as well as political consequences, not just the negative ones that the older philosophies of science presumed. The ones with good scientific effects can also make certain democratic ideals attractive; in early modern science they made the equality of observers, the publicity of information, and the equality of apparently different kinds of bodies appear natural, "common sense," and desirable. We shall return below to identify some of the negative scientific as well as political effects created by other political elements in familiar philosophies of science.

First, however, it needs to be noted that what counts as maximally desirable "democratic ideals" is itself a controversial issue.

Standards for Democratic Social Relations

There will be scientific standards for adopting democratic social ideals into sciences' cognitive cores: such ideals must promise to advance the growth of scientific knowledge (as the familiar ones identified above do) in addition to whatever political benefits they deliver. But what political standards should one use for selecting such democratic social ideals? One approach would be to identify a (hopefully relatively uncontroversial) general democratic principle, and then try to specify social practices that would conform to it. For such a principle one could take the familiar claim that those who bear the consequences of decisions should have proportionate shares in making them. There will be exceptions made to such a general rule, of course: infants and other very carefully identified groups cannot be expected to be able to make such decisions, or to make them wisely. Specification of the exceptions will itself be an important part of any democratic process. Nevertheless, the general principle is attractive because it has guided so many different kinds of effectively democratic practices. The institutions and procedures through which those proportionate shares would be exercised can be expected to differ in different social contexts: practices appropriate for small, homogeneous, and/or oral cultures would have to be different than those appropriate for large, heterogeneous, and multiply literate cultures.

Political philosophers point to the varying democratic effects of three kinds of more specific principles that have been thought to conform to such a democratic directive. One recommends that the interests of relevant groups should be fairly represented during decision-making processes. In local, national, and trans-national scientific councils, the interests of all the groups who will bear the consequences of scientific and technological decisions should be represented during policy-making processes. While this conception of democratic practices is better than none, critics argue that it does not produce democratic enough effects. Who is to represent such interests? (Should those in power be presumed to be able to represent fairly the interests of those over whom they exercise power?) And how are such councils to be held accountable for identifying who the relevant groups are, how their interests can best be fairly represented, and how democratically to resolve conflicts between competing interests? This principle has been used to justify patently unjust social and political systems; it has often been accused of a paternalism which blocks recognition and appropriate consideration of the interests of the politically and socially least powerful groups.

A stronger proposal is that members of such relevant interest groups should themselves have rights to represent their group's interests in decision-making councils: there should be proportionate numbers of women and men, whites and Blacks, and so on among the groups that design and manage scientific institutions and projects. This practice can go far to correct anti-democratic tendencies in any project. There is ample evidence that not only scientific and technological benefits but also political ones can develop from grassroots organization, "participatory action research," "bottom-up" design, and other such ways of giving "end users" a central voice in the design of scientific and technological projects. Nevertheless, these approaches can be enacted in too conservative ways that reduce both their scientific and political benefits. Widespread experience with "adding women and minorities" to worksites and policy groups where they have heretofore been excluded reveals such limitations. Who decides which groups have a right to be so represented? Will only the least threatening members of minoritized groups be the ones permitted into science policy councils? Must they suppress their "difference"—mute their demands or present only those parts of them that easily fit into prevailing concepts and practices—in order to function effectively within the prevailing standards for organizational behavior in scientific and technology institutions? Can individuals produce powerful enough discourses on behalf of their groups' interests to compete effectively with the prevailing dominant discourses, institutions, and practices of the majorities? After all, it took decades of political and intellectual work by many diverse groups to produce the evidence and arguments demonstrating that so-called development was primarily delivering benefits to already advantaged groups and dedeveloping and maldeveloping the already least advantaged. This understanding was not one that automatically became visible to those initially disadvantaged by development policies.

The strongest proposal for achieving democratic standards is that the latter require real equality among groups in the institutions and societies in which such decisions occur. Until sexism, racism, and class systems no longer are able to distribute social, political and economic resources inequitably, institutions within such societies cannot achieve maximally democratic decision processes. "Real equality," moreover, includes symbolic as well as "material" resources. Thus if a culture's ethnocentric standards—the inequality of groups, their thinking, and traditions—are modelled as ideal in the cognitive cores of sciences, one should not expect real equality among members of such groups in scientific institutions and cultures. Insofar as the politics of the larger society are modelled in a science's cognitive core, they serve as obstacles to efforts to eliminate other ways in which such standards shape scientific practices. Attempts at more democratic overt decision processes in scientific institutions are frustrated by models of ideal scientific practices that broadcast antidemocratic messages. The latter are all the more powerful when they are obscured in a cognitive, technical core of science claimed to be immunized against the possibility of social influence and, thus, of critical social analysis.

"More science and technology" in undemocratic societies cannot be expected to deliver the benefits and costs of research democratically, according to this view. In such conditions, more science and technology is guaranteed to increase social inequality (in spite of the intentions of individual scientists or policy makers). Scientific institutions, their cultures, and practices cannot by themselves counter antidemocratic power distributions within society's other institutions that insure that only those already politically, economically and socially advantaged will be in positions to be able to take advantage of the information scientific and technological work produces. When scientific institutions, cultures, and practices produce conflicting political messages, it is even more unreasonable to expect them to be advancing democratic social relations. There are scientific costs to such antidemocratic conditions also, to which the following sections turn. This situation predicts a rather depressing scenario to most of us who thought that more science and technology always delivered at least some benefits "to humanity," however much other benefits were siphoned off by the already overadvantaged.

The next section turns to review one familiar element of modern philosophies of science which, it has been argued, encodes antidemocratic values and interests—namely, the universality ideal. What is antidemocratic about this ideal, how is it politically and scientifically dysfunctional, and what justification is there for nevertheless retrieving a part of it for pro-democratic projects?

One World, One Truth, One Science?

The "unity of science" thesis is one important configuration of ideas in which the universality ideal has played a central role. This model of the sciences and the world they would describe and explain became popular in the late nineteenth and early twentieth centuries; it became a kind of intellectual movement in the first half of the twentieth century. In modified versions it still holds great appeal for many philosophers, scientists, and the general public. However, today it also is the object of considerable skepticism on the part of philosophers and historians of science (Dupre, 1993; Galison & Stump, 1996).

According to this argument, there is one world, one and only one possible true account of it, and one unique science that can capture that one truth most accurately reflecting nature's own order. Less visible in most articulations of the unity thesis is a fourth assumption: there is just one group of humans, one cultural model of the ideal human, to whom nature's true order could become evident.[1] For early modern scientists and philosophers, the ideal human knower could be found among members of the new educated classes. Such individuals could use distinctive knowledge-seeking procedures, and thereby their theories could come to reflect the true order of nature that God's mind had created, just as the latter had also created human minds "in his own image." As the ideal human mind came to occupy the place in modern philosophy that the soul had occupied in Christian thought, rational man replaced spiritual man as the chosen recipient of the one true vision of the world's order.

There are a number of historical problems with the unity of science thesis and its universal science ideal. For one thing, modern science is plural.

There are many distinctive modern sciences with incompatible ontologies, methods, and models of nature and of the research process. If "unity" means singularity rather than simply harmony, the possibility of "reducing" them to one—methodologically, ontologically, theoretically, linguistically—no longer exerts the attraction it did earlier in the century (Hacking, 1996). For example, no longer does it seem reasonable to most philosophers or scientists to try to explain the phenomena of interest to biology referring only to the kinds of natural objects usefully invoked in physics, or by using only the methods, models, or languages useful in physics. Nor are the methods of the sciences reasonably regarded as singular in any interesting sense.

This is not to deny that diverse sciences can usefully share some of their research techniques, models, languages, and other elements. Such borrowings have been a continual source of new insights in every field. Nor is it to deny that they find effective ways to communicate across their differences (Pickering, 1992). Perhaps "unity" should be taken to mean only harmony, as some early defenders of the unity thesis had in mind. However, while many kinds of such harmony certainly do exist among the sciences, such an interpretation of unity undercuts attempts to claim universality for elements of the cognitive cores of sciences. There can be all kinds of "harmonies"—sharings, borrowings, communications—of disparate elements without any elements at all achieving universality.

Moreover, globally, there have been many effective scientific and technological traditions. Elements of these traditions have been borrowed into European sciences, and vice versa (Goonatilake, 1984; Needham, 1954ff; Petitjean, Jami, & Moulin, 1992; Sabra, 1976). Why should anyone, from any culture, value having access to one and only one scientific tradition, as the unity (singularity) of science thesis does, and thereby lose the resources provided by the availability of several or a multiplicity of such traditions? Perhaps there are good reasons for such a value system, but it is worth reflecting on how cognitive diversity in human knowledge-traditions arises, and what is valuable about such diversity.

Sources of Cognitive Diversity

It is easy to see why diversity in science and technology traditions will arise, but it is also not difficult to see why such diversity is a valuable resource for the human scientific and technological legacy (Harding 1998). For one thing, cultures occupy different spatiotemporal locations in nature's heterogeneous order. Some peoples live on deserts, others on fertile plains; some at high altitudes and others at sea level; in warm or in cold climates; in sparse and fragile environments or seemingly bountiful and sturdy ones. Some interact with nature on the sea route from Genoa to the Caribbean; others on the air route from Cape Kennedy to the moon. Each culture that is to survive will have to ask questions relevant to its survival about the environment within which it is located or through which it travels.

Moreover cultures have different interests in the environments with which they interact. Even what is apparently the same environment—a desert, for example—can be the object of different questions by different cultures. Some will want to know how to navigate trade routes across the sands, where camels and their riders can find water, and how both can survive sand storms. Others will want to know how to divert nearby rivers to irrigate a desert for human settlements and even for large-scale agribusiness. Yet other cultures will want information about the geological formations and human populations in particular deserts in order to use these areas as nuclear weapons test ranges. Other cultures will want to know how to mine the minerals and oil that lie beneath a desert's surface. Cultures' different locations in nature's order and their different interests in their environments will lead them to ask different questions and to develop different repositories of knowledge about nature's order. Since a culture's preoccupation with one set of environmental issues can lead it to ignore others, bodies of systematic knowledge are always accompanied by bodies of systematic ignorance: the two are always coproduced.

So far, none of these ways that cultures tend to produce different bodies of knowledge and ignorance would appear to challenge the singularity of science thesis and its universality claim. After all, it is precisely the project of the sciences, the singularity thesis assumes, to try to fit such diverse collections of knowledge into one picture frame that provides the one truth about nature's singular order. The model invoked here is of a jigsaw puzzle in which different sciences, scientific eras, or cultures, contribute different pieces of detail to a more and more complete representation. Yet the next two sources of cognitive diversity show how merely analytic was extraction of the first two sources from the social contexts in which they occur. In reality, these first two apparently untroublesome aspects of cultural difference for the singularity thesis defenders turn out to be inextricable from other kinds of incompatible elements of knowledge systems. These incompatible elements create no insurmountable obstacles to harmony between knowledge-systems (sharings, borrowings, communications), but they block the possibility—and the desirability—of singularity.

What parts of nature it is that cultures perceive as desirable to occupy, how they conceptualize their interests, and just what questions they do ask are shaped also by their local discursive resources—the metaphors, models and narratives with which they have come to understand themselves and their environments. For example, consider the different kinds of questions Europeans have been lead to ask about their environments over the last five centuries as they conceptualized the world around them. The model of nature as an alive organism ("Mother Earth"), who made available an endless cornucopia of resources, directed scientific research. So too did the Christian conception of nature as a product of God's mind; to practice modern scientific methods provided the opportunity to get to know God's mind in greater detail—an opportunity uniquely available to humans, in whose image God had created them, too. At the same time, scientific work was also directed by the representation of nature as a simple mechanism such as a clock. Passing over the next several centuries, we see today representations of nature's order as a complex mechanism such as a biofeedback computer, and in environmental sciences, of the earth as lifeboat or spaceship.

These models of nature drew scientists' attention to different aspects of nature's order, suggested how fruitfully to enlarge the domain to which their theories could apply, and provided ways to reorder familiar "facts" into more satisfying explanatory patterns. Of course most of medieval observations of the movements of comets, planets, and stars are retained in astronomy today. But the relations between these movements, how they should be explained, and what they mean to people have shifted in light of subsequent scientific theories and the social projects with which they have co-emerged and have been co-constituted. Many of "the facts" are indeed always and everywhere the same, but their significance for scientific and social thought can vastly change. Significant here is that the image of the jigsaw puzzle obscures important aspects of this history. The facts about nature's bountiful order produced within early modern scientific models of nature conflict with those produced by contemporary environmental research. Such conflicting bodies of knowledge cannot be fit neatly together to create a singular scientific picture of nature's order.

Finally, cultures organize differently the production of knowledge about the world around them, and how they do so effects what they can know. This is the fundamental principle of scientific method, of course: different kinds of interactions with nature will produce different bodies of knowledge. In this context, the notion of "method" is usefully expanded to include any distinctive way a group organizes its collection of the information it needs to interact effectively with its environment. Oceanographers and climatologists today produce distinctive bodies of knowledge through their research projects. Careful observation of stars, winds, cloud, and wave patterns enabled Pacific islanders to make voyages in open canoes over thousands of miles of open ocean to New Zealand and Australia, for example, and to return safely home (Watson-Verran & Turnbull, 1995). Arctic inhabitants have developed similar knowledge for navigating snow terrains.

Thus, whichever cultural groups have the power and resources to command searches for the kinds of knowledge they want for their projects will develop distinctive repositories of knowledge. These repositories must continually be revised to adjust to changing environmental conditions, new social interests, and the cognitive resources they gain in exchanges with other cultures and their knowledge systems. Moreover, each such body of knowledge, whether in modern science or other knowledge systems, is accompanied by a matching systematic body of ignorance. In choosing to focus on one set of patterns of nature's regularities, with one set of interests, discourses, and methods of knowledge production, a culture leaves unexamined other patterns and the ways of thinking about them that other interests, discourses, and methods could produce. As postcolonial science and technology theorists often ask, what would modern sciences look like if they had been developed in other parts of the world—China, India, the Middle East or Africa—with the differing interests, discourses, and characteristic ways of organizing human activities found in those local cultures instead of in expansionist Europe? We cannot know the answer to such a question, but contemplating it helps one to appreciate the resources provided by these multiple sources of cognitive diversity.

Against such a background, it is easier to appreciate the political and scientific dysfunctionality of the universality ideal.

Costs of the Universality Ideal

We are now in a position to count up the political and scientific costs of the universality ideal.

Political Costs

Arguments in the preceding sections have already pointed to some of the most important political costs of the universality ideal. It supports the devaluation of forms of knowledge-seeking that have proved valuable in other cultures; indeed, of ones that today are crucial to the survival of groups effectively delinked from the benefits of international science and technology. Many African and South American cultures, for example, have little or no access to international science and technology, and survive—and in some cases thrive—thanks to the strengths of their local knowledge traditions.

Moreover, to devalue these traditions is to devalue the people and cultures that use them. This legitimates the continuing forcible subjugation of these groups to western projects—military or commercial. Would the North have so few moral qualms about the sacrifice of third world cultures to purported economic progress if it perceived their knowledge systems as valuable elements for now and for the future of the collectivity of human knowledge? From this perspective, modern sciences have ended up, usually unintentionally, as complicitous with some of the worst genocidal social projects in the name of "human" progress. Would this occur as easily if Northern cultures conceptualized many different centers of human cognitive progress rather than only one—their own? The universality thesis legitimates continuing to move access to nature's resources from those who are already the politically and economically most vulnerable to those who are already the best positioned to take advantage of such access.

Furthermore, the universality ideal supports the construction of models of the rational, the objective, the progressive, the civilized, and the admirably human in terms of distance from the non-European, the economically frugal, as well as the feminine. Moreover, it elevates authoritarianism to a social ideal, for it asserts that it is desirable for everyone to acknowledge the legitimacy of one culture's (the "international science culture's) claim to provide the one true account of the world. The authority of the universality ideal is presented as a necessity for the distinctively rational, progressive, civilized, and human. In such respects, the universality ideal is not at all politically neutral.

Yet such political costs are not the only ones exacted by this ideal.

Scientific Costs

To start with, the universality ideal legitimates decreasing cognitive diversity, yet it is just such diversity that has provided continuing resources for the growth of every culture's scientific and technological projects. Without the availability of other systems of knowledge from which to borrow novel understandings of local environments and the resources they can offer, as well as metaphors, models, and narratives of nature and humans' place in nature, new inquiry techniques and ways of organizing the production of knowledge, any knowledge system would be stuck with only what can be generated from within its own "culture." Modern sciences would have been deeply impoverished without the resources gathered into them from the knowledge traditions of the other cultures Europeans encountered. Moreover, we cannot know what knowledge we will need in the future as social and natural environments change, and new needs and desires develop. Different cultures' constantly evolving ways of thinking about nature and social relations will continue to provide valuable resources for each others' projects. It is as foolish to decrease cognitive diversity as it is to decrease biological diversity.

Second, the universality ideal legitimates accepting less-well supported claims over potentially stronger ones in many cases. If the ontology of a claim (the aspects of the world on which it focuses), the methods used to gather evidence for it, or the models and narratives through which it approaches nature do not fit with the one prevailing one, it can be ranked as less probable than a claim with far weaker empirical evidence that is consistent with prevailing scientific models. My use of terms such as "less well supported claims" and "weaker empirical evidence" should not be taken to indicate that I assume universally accepted standards for scientific claims or the infallibility of any empirical reports. My point is that legitimating only one culturally decontextualized set of standards for evaluating evidence can prove to be scientifically costly. Environmental studies that rely exclusively on the analyses of physical sciences cannot recognize as valuable components of "the best explanation" the kinds of analyses that social scientists bring to environmental studies. Of course it is valuable for each culture to "test" the claims of others within the resources of its own knowledge-system. What is problematic is to assume that such a procedure correctly identifies the worth of a claim, rather than only its ability to be confirmed within a favored knowledge-system. There is an important difference between cases where there is overwhelming evidence of a claim's falsity, and where a claim has not yet been fairly tested, or where no adequate explanation is yet available for the effectiveness of a particular kind of intervention in natural or social orders.

Thus, in the third place, the universality ideal legitimates resistance to some of the deepest and most telling criticisms of particular scientific claims. Criticisms that cannot be recognized as coming from within established boundaries of scientific discussion can legitimately be devalued or ignored. Thus feminist analyses are persistently conceptualized as coming from outside science, even when the critics are respected scientists as, for example, with feminist biologists' criticisms of standard interpretations of evolutionary theory and medical representations of women's body processes. Similarly, postcolonial criticisms of western scientific and technical expertise is often rejected as coming from outside science even when the critics are trained in western science. This is so even and when the critics' goal is not to reject western scientific expertise wholesale, but rather to better integrate it with insights from local knowledge systems. The ability to detect "rigorous refutations" weakens when rigor is presumed to be the monopoly of the one and only real science.

Next, the universality ideal promotes only narrow conceptions of both nature and science. As long as physics is presumed to be the model for all sciences, whether on historical, ontological (e.g., its focus on primary vs. secondary qualities), methodological, or other grounds, other ways of understanding nature's order will be devalued. To mention just one case, it blocks our ability to bring into focus the social elements—institutions, practices, meanings—in what are often presented as merely natural, scientific, and technological changes.

Another limitation is that the ideal of one true science obscures the fact that any system of knowledge will generate systematic patterns of ignorance as well as of knowledge. Every knowledge system has its limits, since its priorities select which aspects of nature's order to study; which questions to ask; which metaphors, models, narratives, and other discursive resources to use; and which ways to organize the production of knowledge. Knowledge systems are like Thomas Kuhn's paradigms in this respect. They can prove illuminating far from their original sites of production, but they all have their limits, and produce recognizably diminishing returns sooner or later.

Finally, such a model for the natural sciences promotes similar problems in the social sciences that model themselves on the natural sciences, such as physicalist psychologies, rational choice theories in economics, political science, and international relations, and positivistic sociologies. Moreover, the prevalence of such models in the social sciences has bad effects in another respect; social scientists who oppose such models often can see no alternative but to focus entirely on the micro and the local. Relativist epistemological positions start to look far too attractive as long as the universality ideal is the only alternative. As a reaction to naturalistic social sciences' devaluations of the local, these other social research projects get contained by the local. The universalist and relativist positions are really two sides to the same coin. In effect, universalism's conceptual world is advanced in unarticulated forms through the relativist positions.

Thus adherence to the universality ideal brings costly political and scientific consequences.[2]

Universalizing Universalizing

Yet the universality ideal captures some features of scientific work that deserve to be preserved, whether in international science cultures or in others. My argument is not to abandon the ideal completely. For one thing, it is too deeply a part of our own western Enlightenment legacy to be abandoned so easily. It is an escapist fantasy to imagine we could accomplish such a feat. Setting out to abandon such a central part of a dominant western self-image assures a "bohemian" status for such a traveller. Instead, my goal is to identify what parts of this ideal are still valuable, and how they could be appropriately reconceptualized.

One useful idea here is that while all beliefs and technological practices are generated in the heat of some culturally local project or other, some prove far more useful than others. Beliefs are certainly not automatically more valuable just because they are "local." After all, individuals and cultures lead nasty, short, and brutish lives if they do not figure out how to avoid damage from excessive cold and heat, hurricanes, floods, fires, and ozone holes, exposure to deadly diseases, or poisoning by nicotine or toxic wastes. Some beliefs travel well, persisting and becoming useful at other places and times, and in contexts very different from those of their origination. So one could say that the attempt to *universalize* a belief is simply the attempt to see in what contexts it can gather empirical evidence and prove useful. Restricting this term to its verbal form simply describes an activity common to any knowledge system. It does not commit us to the claim that there is only one truth about the world and one science that can capture that truth.

Nor does it commit us to the fourth assumption of the unity of science claim: that there is one and only one group or culture of humans that can develop that science. As the histories of science reveal, elements of many different cultures' science and technology traditions have found a home in western—now, "international"—science at one point or another, just as elements of western sciences have been integrated into other knowledge systems. All knowledge systems are hybrids; their ability to continue to grow comes from the access they have to continually new cognitive and material—natural and technological—resources. So this universalizing process can and does occur in many different simultaneously existing knowledge systems. We could conceptualize as valuable the universalizing of different cultures' universalizing practices, not just that this process occurs only in modern western sciences. Important elements of many different cultures' inquiry traditions have become valuable within modern biomedicine's different knowledge system, just as elements of Ptolemaic astronomy remained valuable within the vastly different conceptual world of Copernican astronomy.

Such a perspective could lead to prioritizing the development of significantly different knowledge systems rather than of only one perfect system. And here is just one way that philosophies of modern sciences could encode democratic ideals more effectively with benefits both for democratic social movements and for maintaining the cognitive resources every knowledge system needs to flourish.[3]

Notes

1. Political Philosopher Val Plumwood pointed this out to me.

2. Several of the dysfunctional consequences listed here occur also in John Dupre's (1996) list of problems with what he refers to as the "unity of scientism" thesis.

3. Several themes in this essay have been developed also in Harding, 1998.

Bibliography

Dupre, J. (1993). *The disorder of things: Metaphysical foundations for the disunity of science.* Cambridge: Harvard University Press.

Dupre, J. (1996). Metaphysical disorder and scientific disunity. In P. Galison & D. Stump (Eds.), *The disunity of science.* Stanford: Stanford University Press.

Galison, P., & Stump, D. (Eds.). (1996). *The disunity of science*. Stanford: Stanford University Press.

Goonatilake, S. (1984). *Aborted discovery: Science and creativity in the third world*. London: Zed.

Hacking, I. (1996). The disunities of the sciences. In P. Galison & D. J. Stump (Eds.), *The disunity of science*. Stanford: Stanford University Press.

Harcourt, W. (Ed.). (1994). *Feminist perspectives on sustainable development*. London: Zed.

Harding, S. (1998). *Is science multicultural? Postcolonialisms, feminisms, and epistemologies*. Bloomington: Indiana University Press.

Needham, J. (1954ff). *Science and civilisation in China*. 7 vols. Cambridge: Cambridge University Press.

Petitjean, P., Jami, C., & Moulin, A. M. (Eds.). (1992). *Science and empires: Historical studies about scientific development and European expansion*. Dordrecht: Kluwer.

Pickering, A. (1992). Objectivity and the mangle of practice. *Annals of Scholarship, 8:3*, ed. Alan Megill.

Sabra, I. A. (1976). The scientific enterprise. In B. Lewis, (Ed.), *The world of Islam*. London: Thames and Hudson.

Sachs, W. (Ed.). (1992). *The development dictionary: A guide to knowledge as power*. Atlantic Highlands, NJ: Zed.

Sparr, P. (Ed.). (1994). *Mortgaging women's lives: Feminist critiques of structural adjustment*. London: Zed.

Watson-Verran, H., & Turnbull, D. (1995). Science and other indigenous knowledge systems. In S. Jasanoff, G. Markle, T. Pinch, & J. Petersen Eds.), *Handbook of science and technology studies* (pp. 115–139). Thousand Oaks, CA: Sage.

8

Democratizations of Science and Technology

DANIEL LEE KLEINMAN

Discussions of democratic involvement in science and technology are often marred by murkiness and consequent misunderstanding. In this chapter, I hope to bring some clarity to these often blurry discussions and to make a general case for citizen participation in the realms of science and technology. Toward this end, I do three things. First, drawing on existing case studies and analyses, I distinguish between several forms of citizen participation in matters of science and technology. The exercise in clarification will, I hope, make evident that productive debate on the issues of citizen participation in the realms of science and technology depends on agreement on terms. It is certainly possible for debate participants to oppose any democratic involvement in matters of science and technology; however, if discussants are operating with different definitions then disagreement may be the result of misunderstanding rather than divergent values or assessments. Clarity over the terms of discussion will permit debate to move beyond overblown rhetoric and caricature to specification of substantive differences.

The second task of this chapter is evidentiary. A great deal of opposition to citizen involvement in the realm of science and technology is based on the claim that laypersons are incapable of grasping the technical nuance and methodological complexity of science (cf. Levitt & Gross, 1994b). The case studies I describe suggest that this is not always the case, and consequently, such assertions do not provide an a priori basis for excluding laypeople from decision making about science and technology.[1]

If citizens are capable of intelligent participation in debates about science and technology, the crucial question is, Under what conditions? This is the third matter I consider. I reflect on the cases covered in earlier sections, and I suggest that optimal citizen participation depends on adequate time and other resources, opportunity to examine deeply held assumptions, and mechanisms that weaken the effects of socially significant forms of inequality. In this context, I briefly outline proposals that might enhance democratic involvement in matters of science and technology.[2]

Varieties of Democratized Science

Examples of citizen involvement in the realm of technoscience can be distinguished across several dimensions.[3] First, we must ask what the nature of lay involvement is. To what extent does this participation involve laypersons in activities and decision making commonly understood to be the exclusive realm of experts? At what point do laypeople enter the process? Second, we must understand the nature of expert involvement. Third, we must consider the organizational dynamic of citizen-scientist interaction. Who defines the nature of each group's involvement? Who defines the terms of discussion or participation? In what ways and to what extent do laypeople rely on experts for data, measures, and analysis? Finally, we must assess the extent to which involved actors view "technical" and "nontechnical" (e.g., social and ethical) considerations as discrete and the extent to which they see "technical" matters as appropriate for consideration by laypeople.[4] The classic position in the scientific community is that technical matters are the sole realm of experts and only nontechnical issues are properly the domain of laypeople.[5]

Given these dimensions, it is possible to place examples of citizen involvement in science on a continuum. Such placement is imprecise, in part, because a continuum runs along a single dimension, and here we are dealing with four. This is, nevertheless, not a fruitless exercise, because these four dimensions tend to be highly correlated in actual cases. What is more, placing actual cases along such a continuum and evaluating them across the four dimensions I outline permits us to gain purchase on the distinctive character of each case.

All cases are defined in contrast to what might be termed scientist self-governance. Michael Polanyi nicely characterizes this idea. In such an environment, according to Polanyi, "the choice of subjects and the actual conduct of research is entirely the responsibility of the individual scientist, [and] the recognition of claims to discoveries is under the jurisdiction of scientific opinion expressed by scientists as a body" (1951, p. 53). More generally, this orientation suggests that only experts are competent to make decisions about science and its trajectory, and the "popular will" can play no part in such deliberation (Polanyi, 1962, pp. 67, 72).

At the least contentious end of the continuum would be cases in which scientists acknowledge a social dimension to a problem and all parties agree that this is appropriately a realm for nonscientists, while technical questions are matters for expert consideration. One could place traditional processes for establishing federal research funding priorities here. Elected representatives are the people who determine which federal agencies should get money for research and to which programs this funding should be allocated. Scientists determine what is feasible research and assess the "technical" merit of specific research proposals.[6]

Toward the other end would be cases in which lay citizens challenge the rules of the scientific method and are involved in the production and evaluation of knowledge. Such cases would constitute a radical variety of democratization because laypeople are involved in decision making and other practices that are traditionally viewed as solely the purview of trained scientists. In addition, such cases might spotlight the difficulty in maintaining the technical and nontechnical as discrete realms. Citizens might assert that appropriate research design and data collection procedures must be shaped by nontechnical considerations.[7]

Democratizing Science Policy

I make no claim that the examples of "democratic science" that I profile cover every point on the continuum. Instead, I have selected several instances that are prominent in the existing literature and highlight distinctive and important factors to contemplate in an effort to understand what is at issue in involving citizens in the realm of science.

As I noted, at one endpoint, I place research priority setting in which elected representatives specify social concerns and scientists define technical feasibility and merit. Right next to this instance, we can place the National Institutes of Health's (NIH) efforts to increase citizen input in determining the allocation of resources for scientific research. In the mid-1980s, the NIH initiated a program to include consumers on advisory panels responsible for reviewing research proposals. In these cases, proposal review proceeded in two stages. First, scientists determined the technical merits of proposals and then an advisory council composed of scientists and laypeople decided whether a particular research project should be supported based on the prior scientific assessment as well as NIH priorities and social considerations (Dickson, 1988, p. 327).

I distinguish this case from the role played in a related process by members of Congress because while the Congress has recognized constitutional responsibility for allocating federal resources, lay citizens are only guaranteed this right indirectly through election of their representatives. What is more, while congressional decisions concerning budgetary matters are legally enforceable, the determinations of citizens on these boards were strictly advisory. Interestingly, in this instance, technical criteria were given priority over broadly social considerations. One could imagine reversing the order of assessment. In this case, a lay citizen body would first determine social priorities. Following this, scientists would identify which proposals meeting social criteria were also technically acceptable.

What makes this case among the least radical efforts to democratize science is that it accepts the principle of scientist self-governance, and there is a clear distinction made between the technical merits of proposals—about which only certified experts are assumed able to judge—and the value-laden questions of social priority. The decision-making process is premised on a traditional division of responsibilities in which laypeople deal with nontechnical matters and scientists evaluate technical matters. Furthermore, in terms of processual issues scientists play a role in making decisions about the nontechnical, but the reverse is not the case. Finally, unlike cases that constitute a serious challenge to traditional views of scientist self-governance, it is taken for granted in this case that it is possible to make a clear distinction between the technical and the social, and that it is appropriate to do so.

The controversy in the late 1970s and the early 1980s over the potential hazards of genetic engineering or recombinant DNA (rDNA) research brought the problem of citizen involvement in the regulation of scientific investigation to public attention (Krimsky, 1982; Wright, 1994). In several prominent instances, lay citizens were involved in establishing guidelines for genetic engineering research. Although falling on the continuum close to the case of priority setting just discussed, these instances should be located at a point further from scientist self-governance. On the one hand, in the former case it is widely held to be legitimate for nonscientists to determine the allocation of tax money on the basis of nontechnical criteria. On the other hand, assessment on technical and nontechnical dimensions are seen as discrete acts, and lay citizens are to have no role in technical assessment.

In cases in which citizens are involved in establishing research guidelines, these laypeople must grapple with technical material. Considerations are not restricted to "social" and ethical matters, and indeed, in the two instances I describe below, permitting social considerations to enter discussion is viewed as illegitimate. These cases still exhibit a rather traditional division of labor: scientists determine what counts as risk and laypersons determine what level of risk is acceptable.

At the federal level, in 1974 the National Institutes of Health (NIH) established a Recombinant DNA Advisory Committee (RAC) and charged it with developing and coordinating the implementation of federal guidelines for the conduct of recombinant DNA (rDNA) research. From the perspective of NIH officials responsible for establishing the RAC, it was to be an expert committee (Wright, 1994, p. 165). Indeed, early high-publicity debate about genetic engineering research suggested that matters of social impacts and ethical considerations ought to be set aside and that only scientists were qualified to make judgments about the technical considerations surrounding potential health and environmental hazards (Wright, 1994; Krimsky, 1982).

Public pressure led to the inclusion of one nonscientist on the committee in April of 1976 and a second in September of that year. One analyst suggests that the views of these lay participants were largely marginalized (Wright, 1994), and an NIH public hearing on the regulation of rDNA research came, according to this analyst, "late in the policymaking process, after a consensus on major policy issues had already been established" (Wright, 1994, p. 214). Ultimately, the release of the first formal NIH guidelines on rDNA research in June 1976 reflected the views of the biomedical research community (Krimsky, 1982, p. 154–164; Wright, 1994, p. 190).

When NIH guidelines were subsequently revised by the RAC, political protest led the committee to include a provision in the new guidelines that called for broader membership on the committee. Henceforth, 20 percent of committee membership would be drawn from "persons knowledgeable in such matters as applicable law, standards of professional conduct and practice, public and occupational health, and environmental safety" (RAC document quoted in Wright, 1994, p. 310). Despite this broadening of the committee, "its largest and most influential block could be expected to [support further relaxation of the guidelines], and only a small minority had embraced more cautious positions" (Wright, 1994, p. 354). What is more, a majority of the committee clearly favored absolute self-regulation in research, and only a minority argued in favor of anticipating emergent hazards to workers and the environment (Wright, 1994, p. 356). Finally, even if the committee had not been divided in this way, the NIH director and his allies on the committee set the RAC's agenda, and committee processes were structured to make challenges to the mainstream on the committee difficult.

At one level, as a case of citizen participation in the work of science, the RAC is hardly worthy of mention. Analysts of the RAC have suggested that organizational decisions were made to give the appearance of public participation without permitting citizens to have much real influence (Krimsky, 1982; Dutton, Preston, & Pfund, 1988; Wright, 1994). Indeed, initially the RAC was an expert self-governance body and thus stands beyond the continuum of democratic participation. Still, early on, non-scientists were included on the committee, and thus it constitutes a case, if only a weak one, of citizen participation. To reiterate: it belongs close to the self-governance end of the continuum because it was virtually, although not formally, a case of scientist self-governance. Scientists defined the agenda, marginalized lay input, and excluded non-technical matters from serious consideration.[8]

The Cambridge Massachusetts Laboratory Experimentation Review Board (CERB) (see Krimsky, 1982; Krimsky, 1986a; Goodell, 1979; Lear, 1978; Waddell, 1989; Dutton, Preston, & Pfund 1988) was established, like the RAC, to confront the problem of rDNA research regulation; however, it could be viewed as more democratic than the RAC because it was by design made up exclusively of nonbiologists. The committee's charge, however, was to examine the same limited technical matters as the RAC studied.

In the spring of 1976, a debate began at Harvard University over the renovation of a university laboratory; the lab's purpose was to perform gene splicing experiments of a class designated at the time as "moderately risky." Campus discussions spilled over into the town, and the Cambridge mayor scheduled a hearing on the issue. Ultimately, public concern led the Cambridge City Council to call for a temporary good faith moratorium on high containment rDNA research and to require the establishment of a review board to propose city policy on the regulation of gene splicing research.

The board membership was determined by the Cambridge city manager, who excluded biologists from the committee on the grounds that there was a division within the biology community on the matter and that they had already shown themselves to be an interested party. Board members reflected a diversity of city geography and a range of political positions. The occupations of board members included structural engineer, physician, philosopher of science, nurse/hospital administrator, nurse/social worker, community activist, former city councillor, and businessperson and former mayor.

The board met twice a week for over four months in late 1976, spending approximately 100 hours in session. Using a citizen jury system, roughly one quarter of their time was spent hearing from and asking questions of proponents and opponents of rDNA research. The remainder was spent in group study. In addition, members were expected to read the extensive materials that were distributed and to reach an understanding of the issues. The committee's report, included a draft ordinance slightly stricter than that proposed in the RAC-developed NIH containment guidelines and a call for establishment of a city biohazards committee to monitor local laboratories. Many observers, including scientists, widely acclaimed the process and the final product of the committee's work. A division head at the National Cancer Institute wrote at the time that the committee's report "must surely quiet doubts about the ability of the lay public to deal intelligently and forthrightly with complex technical issues" (Singer, 1977, p. 30; see also Jennings, 1986).

Although this case clearly moves further from the self-governance end of the continuum than the RAC, it does not approach the continuum's other end. Although the committee was made up of nonbiologists, it was entirely dependent on the judgments of scientists. What is more, the committee's mandate, which its members accepted, was defined in terms most congenial to the scientific community. Only matters concerning risks to human health understood as technical would be considered in establishing guidelines.

European "consensus conferences," because they deal equally with the "technical" and "social" matters associated with technological and scientific developments, might be viewed as a further move on the continuum away from scientist self-governance. In addition, unlike the RAC and CERB where the agenda was either directly or indirectly set by scientists, consensus panels, in theory at least, make laypeople central to deliberations, permitting nonexperts to control the agenda, "rather than merely respond[ing] to an already established 'expert' agenda" (Barns, 1995, p. 200).

As Richard Sclove discusses in chapter 2 of this volume, consensus conferences were pioneered in the late 1980s by the Danish Board of Technology, a parliamentary agency charged with assessing technologies. In the last decade the Board has organized a dozen conferences on topics ranging from genetic engineering to the future of private automobiles. Topics are selected on the basis of their broad social and legislative importance, and drawing conclusions demands that conference participants examine diverse dimensions of the issues under consideration.

With a topic chosen, the Board puts together a committee of citizens representing a wide array of social interests and regions of the country to organize the conference. The Board then advertises for volunteer lay participants who provide written statements of interest and are selected to ensure representation of the social diversity of the nation. Participants have access to information from commissioned expert background papers, meetings with "expert panels" representing a wide range of viewpoints, and public fora. Ultimately, the group meets to draw conclusions on the basis of the information it has reviewed.

According to one analysis of a British consensus conference "Despite having no prior background knowledge in the area, [conference participants] . . . developed a degree of understanding and control of the range of complex issues, a competence which was particularly evident in the way they handled the 'experts' and in the production of a very impressive, lucid, and coherent report under rather tight and stressful circumstances" (Barns, 1995, p. 202).

While clearly a further step away from scientist self-governance than the various genetic engineering regulatory bodies, lay conferees are nevertheless largely dependent upon information provided to them by certified experts. In the case of one British consensus conference, Ian Barns goes even further suggesting that "it was largely taken for granted [by conference participants] that the task of technology assessment depended primarily upon the professional and technical skills of research scientists." According to Barns, this assumption "effectively maintained a clear hierarchical difference between 'lay' and 'expert' knowledges" (Barns, 1995, p. 203).

Democratizing Knowledge Production

Moving further along the continuum away from scientist self-governance is a practice Phil Brown and Edwin Mikkelsen (1990) call "popular epidemiology." Epidemiology is concerned with the distribution of disease or a physical condition and the factors that explain this distribution. Whereas the techniques for studying epidemiological phenomena were developed by trained scientists, popular epidemiology "is a process by which laypersons gather scientific data and other information and direct and marshal the knowledge and resources of experts to understand the epidemiology of disease" (Brown & Mikkelsen, 1990, pp. 125–126).

Living in a community where hazards are present often gives citizens access to information about themselves and their environment before it is visible to outsiders, including scientists. Observations made in the process of carrying out everyday responsibilities can lead residents to develop hypotheses about the relationship between incidence of disease in their community and pollutants of some sort. Residents may then carry out their own study or work with trained researchers to execute it.

In Woburn, Massachusetts, for example, local residents were the first to notice a leukemia cluster in their area (see Brown & Mikkelsen, 1990; see also Krimsky, 1984a, pp. 253–255). Residents hypothesized a link between local water (which they noticed was foul smelling and left stains in their sinks) and disease and prodded government officials to undertake tests. Tests found carcinogens present in the water, and these tests and a community health survey in which residents collaborated with biostatisticians at Harvard confirmed residents' hypothesis of a link between the water and cancer. In preparing the survey instrument and the survey design, scholars taught citizens about statistical and epidemiological methods, and residents provided advice on lay phrasing of questions and pointed to additional issues for investigation.

Popular epidemiology differs from traditional epidemiology in its emphasis on "social structural factors as part of the causative chain of the disease" (Brown & Mikkelsen, 1990, p. 126). Concretely, this means that local residents are more likely than traditional epidemiologists to be attentive to factors of politics and economy (e.g., the absence of adequate incentives to push corporations away from polluting, the intimate relationship that can exist between government regulators and corporate polluters, and the inadequacy of government monitoring budgets) and to call for changes in corporate practice and government policies in response to epidemiological findings.

Given the stakes in their communities, advocates of popular epidemiology would rather claim an association between variables when there is not one than to mistakenly overlook an association where there is one. In the terms of traditional research method, popular epidemiologists prefer false positives (or type one errors) to false negatives (type two errors). Professional epidemiologists prefer type two errors, and this preference points to their particular stake in the research process. As Beverley Paigen of the New York State Department of Health notes:

> The degree to which one is willing to make one or the other kind of error is a value judgement and depends on what one perceives to be the consequences of making the error. To conclude that something is real when it is not means that a scientist has followed a false lead or published a paper that later turns out to be incorrect. This may be embarrassing and harmful to a scientist's reputation. In contrast, to ignore the existence of something real means that a scientist fails to make a discovery. This may be disappointing but it does not harm the scientist's reputation, so the scientist is more willing to make type II errors. (quoted in Brown and Mikkelsen 1990, p. 126)

Popular epidemiology falls near the end of the democratic science continuum opposite scientist self-governance for two reasons. First, laypeople are engaged in practices typically reserved for certified scientists. They are not only making decisions about social priorities or defining what constitutes socially acceptable risks, but they are engaged in hypothesis formation, research design, data collection, and data analysis. Second, popular epidemiology challenges the very idea that a hermetic boundary can be established between the technical and the nontechnical. Practitioners of popular epidemiology have challenged the assumption that research design and data collection and analysis in traditional epidemiology occur unaffected by the prior assumptions and commitments of practitioners. In addition, citizens engaged in popular epidemiological work often have very personal stakes in the work they undertake, and they acknowledge that this investment influences the questions they ask and the standards of certainty they find acceptable.

These efforts do not exist at the extreme opposite end of the continuum, where "knowledge" is produced by laypeople with no connection to the work of certified scientists. Citizens engaged in popular epidemiology typically cooperate with traditional epidemiologists. And although the knowledge production practices of these laypeople differ from those of traditional epidemiologists, they are generally consistent with the standards of traditional epidemiology.

Like popular epidemiology, the role of AIDS treatment activists in the production and evaluation of biomedical knowledge is an example of democratized science, far removed from the scientist self-governance end of the continuum. As described by Steven Epstein in chapter 1 of this collection, AIDS treatment activists have been involved in practices traditionally restricted to certified scientists (designing experiments, collecting data, etc.), sometimes with the cooperation of scientists and sometimes not. In addition, activists have sometimes prevailed in arguments with scientists that the technical and nontechnical—here the scientific and the ethical—are not easily separated.

By the mid-1980s, activists were becoming increasingly frustrated by the approval rate of experimental AIDS treatments and the "pace and scope of mainstream research" (Indyk & Rier, 1993, p. 6; see also Epstein, 1995). They became vocal in their criticisms of traditional clinical research. In one instance, activists argued that use of placebos in the Phase II AZT trial was ethically questionable since "in order to be successful the study required that a sufficient number of patients die: only by pointing to deaths in the placebo group could researchers establish that those receiving the active treatment did comparatively better" (Epstein, 1996, p. 202). Activists recommended comparing treatment groups with medical records of matched cohorts of other AIDS patients. Alternatively, patients in the treatment group could be compared with their own medical records from the period before the trial. These kinds of practices had been used in clinical trials in other areas of biomedicine.

This argument went beyond questioning the ethics of using placebos. Activists asserted that clinical subjects concerned about receiving the placebo would find means of obtaining the drug and consequently the "purity" of the control would be undermined. Activists' insights into the kind of trials that would gain the support of people with HIV or AIDS prompted many researchers—biostatisticians in particular—to respect the activists and led activists to have an increasingly important role in discussions about clinical trial design (Epstein, 1996, p. 249).

Beyond pushing for changes in research protocols, treatment activists worked with community medical professionals to design community-based drug trials. Following in a tradition established in cancer research, the County Community Consortium in the San Francisco area gradually became a mechanism for organizing community-based trials. As Epstein relates, "The idea was that physicians would distribute drugs, monitor patients, and collect data as an integral part of their regular clinical work with patients" (1996, pp. 216–217).

The work of the Community Research Initiative (CRI) in New York represents a distinct variant of a community trials model. In this program, people with AIDS or HIV infection participated in decision making about which trials should be conducted and how they should be designed. Drug companies became interested in CRI and signed several contracts with the group to undertake community-based studies (Epstein, 1996, p. 217). Significantly, the U.S. Food and Drug Administration (FDA) relied on data collected in Community Research Initiative and County Community Consortium trials in deciding to approve the drug pentamidine. The commissioner of the FDA praised the CRI trial model. This model has since been used in some trials sponsored by the National Institutes of Health. Importantly, however, this was the first time in the agency's history that it had approved a drug based solely on data from community-based trials (Epstein, 1996, p. 218).

Central to the success of AIDS treatment activists has been the acquisition of a working knowledge of the language and culture of medical science (Epstein, 1995, p. 417). As Epstein describes in his chapter in this collection, many of these activists started with little background in science, but managed to learn the rudiments of AIDS-related biomedicine. Coming to understand AIDS research and clinical practice allowed treatment activists to gain the respect of experts in the field (Epstein, 1995, p. 419; Epstein, 1996, pp. 230–232).

Treatment activists have been successful in challenging the notion that only certified experts can engage in the day-to-day research practices of biomedical science.[9] Their experience provides evidence that it is possible to become conversant in the mode of reasoning and the language of clinical practice without becoming a certified scientist. These activists argued, furthermore, that it is problematic to sharply divide the technical (research methods) from the nontechnical (questions of ethics), and their efforts show that paying attention to the blurred boundary between the two can lead to "better science."[10]

Barriers to Democratizing Science

The cases that I have described suggest that the primary argument used by opponents of democratized science—that laypeople are incapable of grasping the complicated technical material that must be considered (Levitt & Gross, 1994b)—ought not be accepted a priori.[11] However, even if laypeople are potentially capable of intelligent participation in matters often restricted to experts, the barriers to truly democratized science and technology are formidable.

These barriers are part and parcel of the social organization of the United States. It is a society characterized by wide ranging varieties of social and economic inequality and inequity and dominated by the widespread belief in the superior judgment of certified experts.[12] Within this context, citizen participation will often be constrained by the free time (Krimsky, 1984b, p. 48) and economic resources to which citizens have access (Nelkin, 1984, p. 34). In addition, in deliberative bodies composed of lay citizens, social dynamics rooted in such forces as gender inequality are likely to mar the deliberative process (Bohman, 1996).

The case considered above involving the Cambridge Laboratory Experimentation Review Board (CERB) illustrates these kinds of barriers. To begin with, board members accepted the narrowly defined contours of their charge. Following traditional views of expertise (cf. Goggin, 1986b, p. 264; Kleinman & Kloppenburg, 1991), they accepted the sharp distinction between clearly technical and nontechnical issues. They dealt only with the issues surrounding the safety of various genetically engineered organisms and the appropriate physical structures for containing them. What board members and observers understood as ethical and social issues were excluded from formal discussion. This sharp distinction was precisely the outcome that leading scientists in the field wanted. Biologist and active genetic engineering debate participant David Baltimore, for example, talked about leaving questions "replete with value and political motivations" out of the discussion (quoted in Krimsky, 1982, p. 106). This suggests that there is some realm that is free of value and political motivations. But certainly the decision to restrict the discussion is itself "replete with value and political motivations." What is more, decisions about acceptable levels of risk and the balancing of risks and benefits must inevitably be value-laden (cf. Krimsky, 1986b). But these issues were not considered.

In addition to accepting the terms of debate as defined by scientists, the establishment of the CERB reinforced rather than challenged the notion of expertise itself. The board listened almost exclusively to the testimony of certified experts. This procedural decision could certainly have shaped the findings of the board, but in addition, in their own thinking, board members did not escape commonly accepted notions of expertise. Importantly, as board members acknowledged, their decisions were not only influenced by the *substance* of what testifying experts said, but also in important ways by the experts' *credentials* (Goodell, 1979, p. 40).

To suggest that laypersons acknowledged the expert qualifications of some persons and unquestioningly accepted the validity of their claims on that account is not to suggest that the word of experts must inevitably be rejected. Indeed, as Steven Shapin's work (1994) makes clear *trust* is the foundation on which knowledge exists. We must rely on the "word" of others. In securing knowledge, Shapin suggests, "we rely upon others, and we cannot dispense with that reliance. That means that the relations in which we have and hold our knowledge have a moral character, and the word . . . to indicate that moral relation is trust" (1994, p. xxv). Thus, we can only have reliable knowledge to the extent that the people on whom we rely are "reputable and veracious sources, and act appropriately with respect to their testimony" (Shapin, 1994, p. 9).

In the case of scientists, we are asked to trust the institution that they represent and to do so because the norms that govern it are said to stifle regular transgression. But perhaps simple and clear normative transgression is not the problem. In antiquity, "One's word was one's bond only if one was not bound in giving it. The forgoing of free action was considered effective and reliable only if that course was freely decided upon" (Shapin, 1994, p. 39). In the cases under consideration, blanket acquiescence is problematic not because self-policing is inadequate, but because in a world of institutions one's "word" is never "freely" given. As my discussions of popular epidemiology and AIDS treatment activism suggest what counts as a significant finding and what constitutes a legitimate research protocol are shaped by specific institutional histories and, indeed, reinforced by the very norms that are supposed to provide the basis for lay confidence in scientists' words.[13]

Of course, the costs of consistent and unrelenting skepticism would make one's everyday life unbearably difficult. Certainly, most of the time unexamined trust is practically appropriate. But where the future of one's community, family, or person is at issue some measure of skepticism may be healthy. Prodding one's doctor about a diagnosis, and seeking a second and perhaps third opinion, seems entirely reasonable behavior. It is no less the case that citizens should not assume that the way a scientist frames a problem or interprets data is valid and appropriate merely because of the credentials the scientist holds.

Beyond problems of the status of outside experts, within the board itself a power dynamic mirrored those existing in the broader society. The Cambridge city manager selected board members in part on the basis of their own expertise. Thus, the CERB included medical professionals who in theory could speak to issues of health hazard and an engineer who could evaluate the structural efficacy of proposed containment facilities. This decision was reinforced by board members' own attitudes toward other board members. Board members took for granted the legitimacy of generally accepted boundaries between lay and expert realms and granted traditionally defined experts privileged status. According to one analyst of the body's history, the confidence of board members varied enormously, "ranging from two physicians who quickly became vocal proponents of the research, and . . . a philosopher of science from Tufts who played 'devils advocate,' to a nurse and a nun who hardly spoke at all." Everyone on the CERB, according to this analyst, "looked to the two physicians for arbitration of technical difficulties and to [the philosopher] . . . to find any chinks in their armor" (Goodell, 1979, p. 39). In his own discussion of the board, former board member, the philosopher, Sheldon Krimsky suggested that the powerful influence of the "medical people on the committee was justified by their knowledge" (Krimsky, 1982, p. 302). Here, Krimsky seems to be taking for granted the appropriateness of the social status of the medical profession and ignoring the possibility that even an expert's evaluation of information *may* be affected by an array of interests (e.g., in the unambiguous benefits of scientific research), beliefs (e.g., the relative infallibility of doctors), and values (e.g., the appropriate balance between risk and benefit). In addition, the internal dynamic of the committee may not have been exclusively shaped by the social status granted experts. Research suggests that the most vocal board members were men, and that the women spoke rarely (Goodell, 1979). Thus, a gender dynamic may have been at work.

Truly open—and hence democratic—debate was hindered in the Cambridge case by two other factors. First, no effort was made by those with knowledge considered relevant to the committee's charge to systematically share that knowledge with others on the committee (Krimsky, 1982, p. 302). Second, citizens seeking to obtain a comprehensive picture of the problem under consideration confronted a common obstacle: dissenting scientists had a disincentive to offer their perspectives. Goodell (1979) found scientists in the CERB case feared alienating colleagues. Beyond the CERB case, Brown and Mikkelsen (1990, p. 139) found that scientists who aided citizens were sometimes punished. Even where there are no serious threats of sanction, there may be no positive incentives either. Working with community groups on issues of practical importance to them is unlikely to enhance an academic scientist's tenure, promotion prospects, or collegial reputation.

At the far end of the continuum, despite what appears to be a more radical instance of democratic science, AIDS treatment activists have also been unable to escape the politics of expertise and the power dynamics that constitute social relations in the United States. As Epstein points out in the first chapter of this volume, the social status of economically well-off gay white men who composed the core of the AIDS treatment activism movement provided the foundation for the respect they were granted by AIDS medical professionals, and among their "constituency" (see also Epstein, 1996, p. 294). As the demographics of AIDS changes and the growth of AIDS slows among the middle-income gay population, while increasing among IV drug users and people of color, it is unlikely that these latter groups will have the same capacity that gay white men have had to enter the world of biomedicine (Epstein, 1995; Epstein, 1991).

Finally, with so few people in the position to become successful treatment activists—in part because they lack time, economic resources, and social status—there is every likelihood that the treatment movement will reproduce within the community of people with AIDS, or more narrowly the AIDS activist community, the lay/expert dynamic that exists in society at large (Epstein, 1991, pp. 52, 53, 60; Epstein, 1996, p. 294). There will be those few who are in a position to enter the "halls of science" and those who must depend on movement activists for information, advice, and representation. Indeed, Epstein suggests that this is already beginning to occur (Epstein, 1996, p. 288). Of course, as other analysts note (Indyk & Rier, 1993, p. 11; Epstein, 1995), the question of representation—closely related to the problem of democracy—raises the important issue of whether middle-class gay white men represent all people with AIDS in any case or only activists. Women and people of color in the AIDS treatment activism movement have been critical of treatment activist leaders for their inattention to issues of concern to "minority" communities among people with AIDS or HIV disease (Epstein, 1996, p. 291).[14]

Strategies for Overcoming the Obstacles

The obstacles to democratizing science within the existing social order are formidable, and some may be ultimately insurmountable. But while the barriers are unlikely to be entirely transcended, there are a range of possible strategies that would increase the likelihood that these obstacles can be at least partially overcome and the quality of outcomes from efforts to democratize science can be enhanced.

Instances of working-class citizen participation in popular epidemiology notwithstanding, lack of resources is an important barrier to broad social representation in efforts to democratize science from citizen advisory boards to AIDS treatment activism. One must be able to afford time away from work and family to participate, and the resource requirement will be significantly greater in cases in which broad technical mastery is necessary (like AIDS treatment activism), in contrast to cases in which citizens must simply attend to the testimony of certified experts.

Participation in juries in civil and criminal trials is widely considered a responsibility of citizenship in the United States, and jurors are granted a per diem. Federal advisory panels similarly offer payment for daily expenses. For jury service, the payment offered is plainly inadequate compensation. The idea, however, is important. For the array of citizen bodies that provide advice to governments, economic leaders, and the public at large, including community boards and experiments with consensus conferences, some type of per diem system could weaken economic barriers to widespread citizen participation. In his chapter, Richard Sclove estimates a national consensus conference in the United States would cost in the neighborhood of $500,000. This is a relatively small portion of our entire federal budget. One could imagine, therefore, that elected representatives could find small amounts of money to initiate experiments in democratic technoscience of various kinds. However, where fiscal constraints or political barriers make government contributions to such a system impossible, support from private foundations with a commitment to democracy and public understanding of public policy and socially relevant science and technology might be a viable alternative to government support.

Cases like the Woburn popular epidemiology effort and the work of AIDS treatment activists require a greater commitment of citizen time and are consequently more costly. As a way of enhancing citizens' "appreciation of the diverse needs of other communities, [and providing them] a broader experiential basis from which to conceive of their society's general interest . . . ," Richard Sclove proposes "citizen sabbaticals." Sclove views these as analogous to faculty sabbaticals or the U.S. Peace Corps. Such sabbaticals would "encourage each person to occasionally take a leave of absence from his or her home community, to live and work for perhaps a month each year or a year each decade in another community, culture, or region" (1995, p. 43).

For the purpose of increasing lay involvement in the production and evaluation of scientific knowledge and technologies, citizen fellowships might be established along similar lines. Again, if government funds were not available, private foundations might be in the position to establish an endowment that would provide funds to allow a limited number of citizens to take leaves of absence from their jobs in order to work for an extended period on a science-related project. Nonprofit organizations doing science- and technology-related work, for-profit companies, and government and university laboratories and science departments could list opportunities in a database compiled by the entity overseeing the endowment. Citizens would then choose one of these opportunities and submit a statement of interest. Statements would be evaluated and citizens selected by a committee composed of people representing diverse interests and social, economic, and professional backgrounds.

The virtue of such a program from the perspective of enhancing democratization is beyond doubt. What citizens and participating organizations would gain is certainly open to question and would vary. One can imagine, however, a case in which a farmer received such a fellowship and went to work in an agricultural biology lab on a nearby university campus. The farmer and the scientists would have an opportunity to come to know one another. Developing rapport and respect, all parties might conceivably leave the project with enhanced empathy for the others' interests and needs. In addition, however, the research subsequently produced by the lab might benefit from the synergy of the intimate knowledge of farming the farmer brought to the lab and the more traditional biological science in which the lab researchers typically work (cf. Krimsky, 1984a). AIDS researcher Anthony Fauci has spoken enthusiastically about the time treatment activists spent in his lab. Activists gained a better understanding of the "bench science," and researchers were forced to confront the realities of the disease. As some scientists have gratefully acknowledged, activist involvement with basic AIDS research also prompted dialogue between scientists in distinct specialties (Epstein, 1996, pp. 321, 322).

If citizen involvement in the realm of science is to be successful, work must be undertaken to institutionalize mechanisms that allow participants the opportunity to acquire the broadest possible "knowledge base," that promote reflection on taken-for-granted attitudes toward expertise held by participants (Laird, 1993, p. 354), and that maximize the possibility of equal roles for all participants. Among the existing mechanisms for promoting equitable participation is the effort undertaken by a Canadian inquiry into the costs and benefits of proposals to mine natural gas in a relatively pristine part of the Canadian landscape. Among other things, inquiry organizers arranged for testimony from nearly a thousand native witnesses near their homes. According to one analyst, "The familiarity of a local setting and the company of family and neighbors encouraged witness spontaneity and frankness" (Sclove, 1995, p. 29).

In a different context, some university educators have worked to promote learning environments in which students are not structurally excluded from participation. For example, a common practice is to break large lecture classes down into smaller discussion groups. These groups work for a period without monitoring by the teacher and later each group reports back to the classroom as a whole. The advantage of such a procedure is that it can encourage students who are fearful of talking before the entire class to express their views. Perhaps a modification of such a practice could work in citizen science and technology boards.

A different exercise, used sometimes in the classroom, might involve role playing. Citizen bodies might collectively outline the range of positions possible on a given issue and then randomly assign group members to argue for each position. This approach has the advantage that group members with superior presentational and rhetorical capacities, more confidence in public speaking, and/or respect based on some credential not automatically warranting respect would not always be arguing for the positions they most favor and thereby dominating group deliberations.[15]

Trained monitors might observe group deliberations and specific times in group meetings might be set aside for collective self-reflection. Monitors would explore group dynamics, determining who dominates discussion and trying to ascertain why. Special attention would be paid to determine whether certain positions are rejected out of hand by some participants and others are accepted without evaluation. Of course, a methodology for monitoring and assessment would need to be developed, but similar practices are undertaken in classrooms throughout the country with the aim of enhancing teacher effectiveness and prompting more equitable student participation (Sadker & Sadker, 1994).

Knowledge acquisition by citizens interested in participation in techno-science raises perhaps the thorniest problems in achieving useful and success-ful involvement. What makes the AIDS treatment activism case especially interesting is that as outsiders these citizens proposed approaches to research that scientists tended to dismiss, overlook, or ignore. They added a perspec-tive that had been missing. Those instances of citizen involvement in techno-science closer to the scientist self-governance end of the spectrum I outlined in this chapter are less likely to produce novel solutions as they are often structured to accept experts' frameworks and limits of time make the devel-opment of critical perspectives less likely. Still, I believe the kinds of exercises I suggest above make the surfacing of novel orientations more probable.

Cases like AIDS treatment activism are considerably more complicated, however. As Epstein points out in his chapter and elsewhere (1996), once citizens acquire a substantial knowledge base, they are likely to begin to think increasingly like members of the scientific mainstream. They no longer bring the insight of an outsider. Overcoming this dilemma is not likely to be easy. One way to realize the greatest benefit from democratic technoscience involving "lay experts"—those citizens without expert credentials, but with substantial "technical knowledge"—might be to establish a mechanism like those I describe above, but in this instance less knowledgeable partici-pants would have an opportunity to challenge the perspective of the lay experts and lay experts would have a structured opportunity to reflect on their perspective and its transformation as they acquired more formal knowledge.

The "remedies" outlined above speak only to barriers to citizen partici-pation in the practices of democratizing science. But, as all successful efforts in this direction—from consensus conferences to popular epidemiology to AIDS research—suggest, cooperation between scientists and laypersons is absolutely essential. Scientists may on occasion enhance their public legitimacy by working with citizens, but there may be disincentives to such cooperation as well. There may be little that can be formally done to protect participating scientists from colleague ostracism. However, including faculty records of cooperation with citizen groups as part of the service component considered in tenure and promotion decisions for university scientists might very well increase scientists' willingness to cooperate and could conceivably offer them a measure of protection from hostile colleagues. Government or foundation funding for citizen-scientist collaborative research would further increase the likelihood of scientist involvement in work with laypeople, especially if this research led to products (e.g., peer reviewed articles) traditionally valued in the scholarly community.

In terms of promoting cooperation between lay citizens and university scientists, there are contemporary and historical legacies to which supporters of democratized science might look. Collaboration between university biologists and industry is increasingly viewed in a favorable light by academic administrators and can even enhance promotion prospects. What is more, "service," while often devalued, is formally considered in promotion decisions at most universities, and land grant universities have a long tradition of outreach to rural communities. Of course, interest in industry involvement in university science is often predicated on the money it will bring to institutions confronting fiscal hard times, and there is typically little financially that citizens can offer universities. However, citizens are voters, and state universities depend on legislators' kindheartedness. Indeed, seeking legislators' favor could be an incentive for university administrators to take cooperation with citizens more seriously in faculty promotion policies, and certainly at federal land grant institutions could be used to alter current outreach practices.

Although I have pointed to some practical strategies for confronting substantial barriers to the democratization of science and technology, I do not mean to imply that my proposals will be easy to implement or that instituting them will lead straightforwardly to an elusion from the barriers to lay-expert cooperation, to "better" science, or to more vibrant democracy. Volumes on democratic theory and practice fill the stacks of university libraries, and I am not attempting to dismiss the arguments that have raged and the dilemmas that have been debated for centuries. If, however, my discussion contributes even marginally to a shift in the terms of dispute, it will have served its purpose.

Conclusion

In this chapter, I have attempted to contribute to dialogue on matters of citizen participation in the world of science and technology, first by distinguishing between several efforts at democratizing science and technology and thus clarifying some of the terms of debate. In addition, I have shown that some forms of citizen involvement in the realm of science are plausible. Finally, I have suggested that the technical character of science and technology is not always an insurmountable barrier to citizen participation, but have pointed to other barriers that must be overcome if citizen participation is to be optimized.

Science and technology have become central features of our social and economic topography, and developments in science and technology will intimately affect the lives of all people in ways large and small for the foreseeable future. This being the case, it is absolutely imperative that we engage in reasoned debate on matters of citizen participation in the realms of science and technology. Clearly focused discussion with agreement on terms should make this possible.

Acknowledgments

This is a revised version of an article that appeared in *Politics and the Life Sciences* (1998, 17, pp. 133–145) as "Beyond the Science Wars: Contemplating the Democratization of Science." I would like to thank Susan Bernstein, Lawrence Cohen, Scott Frickel, Stuart Greene, Gary Johnson, Gerald Kleinman, Steven Vallas, and the reviewers for *Politics and the Life Sciences* and SUNY Press for their helpful comments and suggestions on earlier versions of this chapter. I benefited greatly from the research assistance of Jana Lonberger.

Notes

1. While I take seriously the importance of lay understanding of the "technical" matters many scientists see as the heart of their work, I agree with critics of traditional notions of "scientific literacy" that our definition of this concept needs to be broadened (Claeson, Martin, Richardson, Schoch-Spana, & Taussig, 1996, p. 102). As Wynne puts it:

> The issues and problems of public understanding of science cannot . . . be divorced . . . from the epistemological issues of the social purposes of knowledge, and what counts as "sound knowledge" for different contexts. These in turn highlight questions about the institutions of science—its forms of ownership, control, and prac-tice. To preclude these issues from public debate is to undermine the possibilities of effective public uptake and culture of science. (1996b, pp. 43, 44)

2. One question I do not confront in this chapter is why, even if demo-cratic involvement is possible, it should be promoted. There is no one answer to this question. A technocratic argument might be that there is no point in increasing citizen involvement in the realm of science and technology, unless the results of such involvement are demonstrably superior to those produced if decision making is restricted to experts alone (cf. Breyer, 1993). The litera-ture on citizen involvement is not univocal on this consideration; however, work by Krimsky (1984), Epstein (1996), and others does suggest that under certain circumstances lay involvement has the potential to produce "better science."

Perhaps the most common argument in favor of democratic involve-ment is that lay citizens have a right to participate in decisions about scien-tific research "which is financed with taxpayer dollars and which has broad social impact" (Goggin, 1984, p. 29). This argument has a lineage dating to the American revolution. A more radical justification for citizen involvement in the realm of science and technology is made by "strong democrats" who argue that "people should be able to influence the basic social circumstances of their lives[,]" which include matters of science and technology (Sclove, 1995, p. 25). Strong democrats call for participatory, not merely representative democracy. Of course, one might argue that decision making about such matters as state supported economic development and social welfare is different in important ways from deliberations surrounding state supported science, and consequently decision making about the latter ought to be managed differently. Similarly, one might argue that participatory democratic involvement in matters of science and technology is simply impractical.

Debate on these matters is far beyond the scope of this essay. However, they are certainly not beyond the reach of public discussions on the role of experts in our society and the appropriate nature and extent of citizen involvement in the realm of science and technology. For an outline of different positions on this issue, see Jennings (1986). For more general discussions of citizen participation, public debate, and democratic theory see, among other works, Bachrach (1975), Barber (1984), Bohman (1996), Garson and Smith (1976), Gelhorn (1972), Heberkin (1976), and Patman (1970).

3. I make no claim that these dimensions are exhaustive, only that they enable us to distinguish different cases and understand both the nature of scientists' resistance and the practical barriers to success. For different but related criteria used in assessing and distinguishing between forms of citizen participation see Arnstein (1969), Krimsky (1984b), and Laird (1993).

4. A wide range of work in the social studies of science and technology suggests that the boundary between the technical and nontechnical—the scientific and the social—is not intrinsic or natural, but is the outcome of sociohistorical processes. As a practical matter, in public debate and discussion between scientists, the division is typically assumed to be sharp and obvious. Technical matters are the substantive considerations with which scientists deal in their research and research-related activity. Social matters concern everything else. Among the issues that commonly fall into this latter category are questions of acceptable risk, moral propriety, and the assignment of broad social priorities in the allocation of tax dollars for research. In this essay, I accept the distinctions made by participants in my cases, supplemented by my sense of social common sense on these matters.

5. See, for example, the quotation attributed to David Baltimore later in this chapter.

6. This characterization oversimplifies the process considerably; the budgeting decisions of Congress are made in light of formal testimony provided to them by scientists and other interested parties. Congressional science policy leaders also often have scientists on staff to aid them in their evaluations. In addition, of course, the Congress is presented with a budget by the President, who relies on members of the executive "expert" in science

matters to provide advice on funding priorities. And this characterization does not include the informal advice members of Congress and the executive receive from an array of interested groups and individuals. The point, however, is that in the end elected officials, not assumed to be experts, are expected to make budgeting decisions in terms of *social* priorities, in light, of course, of what "experts" contend is *technically* feasible. Certainly, Congress sometimes makes decisions to provide resources for scientific and technology projects that members of technoscience communities would normally assert should be subject to peer review. As often, this prompts controversy over governmental meddling in technical spheres and cries of pork barrel funding.

7. Some might assert that this continuum does not extend far enough on the most highly democratized end and that because I utilize traditional definitions of science and knowledge and the technical-social dichotomy (see note 5), I am unable to envision more radical options of democratizing knowledge production. Along these lines, critics of my presentation would question the lay-expert knowledge split that I appear to accept and point out that scientific knowledge embodies an array of values, interests, and assumptions. I would agree with these commentators that we lose a great deal when lay/local/indigenous knowledges are denigrated or ignored, and I accept that scientific knowledge is always infused with a range of "non-technical" values, assumptions, and interests (see Harding in this volume). Although my recommendations for reform do not deal directly with these matters, I do believe that, if implemented, they could alter taken-for-granted attitudes on these issues. Furthermore, the limits of my continuum do not preclude experiments aimed at promoting nontraditional epistemologies. For recent work in science studies that confronts these issues, among other work, see Collins and Pinch (1993); Jasanoff, Markle, Petersen, and Pinch (1995); and Wynne (1996a).

8. Work by Ira Carmen (1992, 1993) suggests that nonbiologists may have had a more substantial role in the recent RAC discussions about human gene therapy than was the case in earlier RAC debates. Whether this means the RAC should now be placed at a different point on the continuum than I am proposing is inconsequential since the examples I outline are intended only as "exemplars." The exercise is not to seek absolute precision in placing all empirical cases along the continuum.

9. Epstein stresses in chapter one of this collection that these activists were by no means typical laypersons. Instead, they were a "new species of expert."

10. As Epstein notes in chapter 1 of this volume, however, the successes of AIDS treatment activists may have unintended consequences as well.

11. To suggest that laypeople are sometimes capable of understanding the technical material that must be considered in decision making about

scientific and technical matters is not to say that lay comprehension is guaranteed in any particular case or that it is easy to come by. Work in "cognitive heuristics" suggests that outside their own areas of expertise, individuals often engage in questionable reasoning, especially on matters of statistics and probability. Indeed, "the tendency to predict the outcome that best represents the data, with insufficient regard to prior probability, has been observed [even] in the intuitive judgements of individuals who have extensive training in statistics" (Tversky & Kahneman, 1982, p. 18). Similarly, Stephen Breyer notes that "People consistently overestimate small probabilities" (1993, p. 36).

Breyer is concerned with federal health and safety regulation and views citizen misunderstanding as a problem because government officials who must make health and safety policy based on risk analyses are not insulated from the citizenry and consequently must make policy influenced by public misunderstanding of risk. The result, according to Breyer, is bad health and safety regulation. The solution, according to Breyer, is not more citizen input, but instead establishment of a centralized government agency that can create a coordinated health and safety policy with a long-term outlook.

Breyer provides an eloquent argument in favor of technocratic policy-making; however, it does not seem to me that the data he marshals illustrates that citizen involvement in risk assessment related policy-making will always result in less than optimal policy. He does not suggest that citizens cannot learn to understand probabilistic reasoning, only that common sense reasoning on probabilistic questions leads to erroneous results. Surely, laypeople can learn probabilistic reasoning, how risk assessment assumptions are made and what the implications of such assumptions are. The question is whether the "benefit" that will result from participation of laypeople with an understanding of risk analysis in health and safety regulation is worth the "cost" of providing an opportunity for lay education. This question is beyond the scope of this chapter. I aim only to indicate that democratic participation in the realm of science and technology is plausible. See also note 3.

It should be noted that there is an extensive literature critical of traditional risk analysis of the kind advocated by Breyer. Wynne, for example, suggests "that public perceptions of and responses to risks are rationally based judgements of the behaviour and trustworthiness of expert institutions, namely those that are supposed to control the risky processes involved" (1996a, p. 57). Following this line of argument, whether citizens understand probabilistic analysis is beside the point. If expert assessments of risk embody implicit "nontechnical" considerations or are simply not truthful, then presented risk probabilities will not be reliable in any case. See also Dutton, Preston, and Pfund (1988, p. 329).

12. Dutton and her colleagues make a related point. They suggest that

Scientific, professional, and corporate groups have a strong vested interest in seeing that innovations go forward, and they typically have substantial financial, organizational, and technical resources to invest in pursuing those interests. Most public groups, by contrast, as well as society as a whole, have a much less direct stake in the outcome of given policy choices, and usually can draw on only limited economic and institutional resources. (1988, pp. 343, 344).

13. In his discussion of sheep farmers and radiation, Wynne (1996a) makes a related point.

14. Beyond these barriers, Epstein points out in his chapter in this volume that "Ironically, insofar as activists start thinking like scientists and not like patients, the grounding for their unique contributions to the science of clinical trials may be in jeopardy" (25).

15. I have borrowed this idea from Steve Schneider who suggested it to me in a conversation over a somewhat different matter: how to make debates over controversies among scientists equitable.

Bibliography

Arnstein, S. (1969, July). A ladder of citizen participation. *American Institute of Planners Journal, 35*(4), 216–224.

Bachrach, P. (1975). Interest, participation and democratic theory. In J. R. Pennock & J. W. Chapman (Eds.), *Public participation in politics*. New York: Lieber-Atherton.

Barber, B. (1984). *Strong democracy: Participatory politics in a new age*. Berkeley: University of California Press.

Barns, I. (1995). Manufacturing consensus?: Reflections on the UK national consensus conference on plant biotechnology. *Science as Culture, 5*, 199–216.

Bijker, W., Hughes, T., & Pinch, T. (Eds.). (1989). *The social construction of technological systems*. Cambridge, MA: MIT Press.

Bohman, J. (1996). *Public deliberation: Pluralism, complexity, and democracy*. Cambridge, MA: MIT Press.

Breyer, S. (1993). *Breaking the vicious circle: Toward effective risk regulation*. Cambridge, MA: Harvard University Press.

Brown, P., & Mikkelsen, E. (1990). *No safe place: Toxic waste, leukemia, and community action*. Berkeley: University of California Press.

Carmen, I. (1992). Debates, divisions, and decisions: Recombinant DNA Advisory Committee authorization of the first human gene transfer experiments. *American Journal of Human Genetics, 50*, 245–260.

Carmen, I. (1993). Human gene therapy: A biopolitical overview and analysis. *Human Gene Therapy, 4,* 187–193.

Claeson, B., Martin, E., Richardson, W., Schoch-Spana, M., & Taussig, K. (1996). Scientific literacy, what it is, why it's important, and why scientists think we don't have it: The case of immunology and the immune system. In L. Nader (Ed.), *Naked science: Anthropological inquiry into boundaries, power, and knowledge* (pp. 101–118). London: Routledge.

Clarke, A., & Fujimura, J. (Eds). (1992). *The right tools for the job: At work in twentieth-century life sciences.* Princeton, NJ: Princeton University Press.

Collins, H., & Pinch, T. (1993). *The golem: What everyone should know about science.* Cambridge: Cambridge University Press.

Cramton, R. (1972). The why, where, and how of broadening public participation in the administrative process. *Georgetown Law Journal, 60,* 3.

Dickson, D. (1988). *The new politics of science.* Chicago: University of Chicago Press.

Dutton, D., with Preston, T., & Pfund, N. (1988). *Worse than the disease: Pitfalls of medical progress.* Cambridge: Cambridge University Press.

Epstein, S. (1991). Democratic science? AIDS activism and the contested construction of knowledge. *Socialist Review, 91,* 35–64.

Epstein, S. (1995). The construction of lay expertise: AIDS activism and the forging of credibility in the reform of clinical trials. *Science, Technology, and Human Values, 20*(4), 408–437.

Epstein, S. (1996). *Impure science: AIDS, activism, and the politics of knowledge.* Berkeley: University of California Press.

Garson, G. D., & Smith, M. P. (Eds.). (1976). *Organizational democracy: Participation and self-management.* Beverly Hills, CA: Sage.

Gelhorn, E. (1972). Public participation in administrative proceedings. *Yale Law Journal, 81,* 67.

Goggin, M. (1984). The life sciences and the public: Is science too important to be left to the scientists? *Politics and the Life Sciences, 3,* 28–40.

Goggin, M. (Ed.). (1986a). *Governing science and technology in a democracy.* Knoxville, TN: University of Tennessee Press.

Goggin, M. (1986b). Governing science and technology: Reconciling science and technology with democracy. In M. Goggin (Ed.), *Governing science and technology in a democracy.* Knoxville, TN: University of Tennessee Press.

Goodell, R. (1979). Public involvement in the DNA controversy: The case of Cambridge, Massachusetts. *Science, Technology, and Human Values, 27,* 36–43.

Heberkin, T. (1976). Some observations on alternative mechanisms for public involvement: The hearing, public opinion poll, the workshop, and the quasi-experiment. *Natural Resources Journal, 16,* 197–212.

Indyk, D., & Rier, D. (1993). Grassroots AIDS knowledge: Implications for the boundaries of science and collective action. *Knowledge, 15,* 3–43.

Jasanoff, S., Markle, G., Petersen, J., & Pinch, T. (Eds.). (1995). *Handbook of science and technology studies.* Thousand Oaks, CA: Sage Publications.

Jennings, B. (1986). Representation and participation in the democratic governance of science and technology. In M. Goggin, (Ed.), *Governing science and technology in a democracy.* Knoxville, TN: University of Tennessee Press.

Kleinman, D. L. (1995, September 29). Why science and scientists are under fire—and how the profession needs to respond. *The Chronicle of Higher Education,* B1–B2.

Kleinman, D., & Kloppenburg, J. (1991). Aiming for the discursive high ground: Monsanto and the biotechnology controversy. *Sociological Forum, 6,* 427–447.

Krimsky, S. (1982). *Genetic alchemy: The social history of the recombinant DNA controversy.* Cambridge, MA: MIT Press.

Krimsky, S. (1984a). Epistemic considerations on the value of folk-wisdom in science and technology. *Policy Studies Review, 3*(2), 246–262.

Krimsky, S. (1984b). Beyond technocracy: New routes for citizen involvement in social risk assessment. In J. Petersen (Ed.), *Citizen participation in science policy* (pp. 43–61). Amherst: University of Massachusetts Press.

Krimsky, S. (1986a). Research under community standards: Three case studies. *Science, Technology, and Human Values, 11,* 14–33.

Krimsky, S. (1986b). Local control of research involving chemical warfare agents. In M. Goggin (Ed.), *Governing science and technology in a democracy.* Knoxville, TN: University of Tennessee Press.

Laird, F. (1993). Participatory analysis, democracy, and technological decision making. *Science, Technology, and Human Values, 18*(3), 341–361.

Latour, B. (1987). *Science in action: How to follow scientists and engineers through society.* Cambridge, MA: Harvard University Press.

Lear, J. (1978). *Recombinant DNA: The untold story.* New York: Crown Publishers.

Levitt, N., & Gross, P. (1994a). *Higher superstition: The academic left and its quarrels with science.* Baltimore, MD: Johns Hopkins University Press.

Levitt, N., & Gross, P. (1994b, October 5). The perils of democratizing science. *The Chronicle of Higher Education,* B1, B2.

Nelkin, D. (1984). Science and technology policy and the democratic process. In James Petersen (Ed.). *Citizen participation in science policy* (pp. 18–39). Amherst: University of Massachusetts Press.

Pateman, C. (1970). *Participation and democratic theory.* Cambridge: Cambridge University Press.

Petersen, J. (Ed.). (1984). *Citizen participation in science policy.* Amherst: University of Massachusetts Press.

Polanyi, M. (1951). *The logic of liberty: Reflections and rejoinders.* Chicago: University of Chicago Press.

Polanyi, M. (1962). The republic of science. *Minerva, 1,* 54–73.

Sadker, M., & Sadker, D. (1994). *Failing at fairness: How our schools cheat girls.* New York: Touchstone.

Sclove, R. (1995). *Democracy and technology.* New York: Guilford Press.

Sclove, R. (1996, July). Town meetings on technology. *Technology Review.*

Shapin, S. (1994). *A social history of truth: Civility and science in seventeenth century England.* Chicago: University of Chicago Press.

Singer, M. (1977). The involvement of scientists. In National Academy of Science, *Research with recombinant DNA: An academy forum* (pp. 24–30). Washington, DC: National Academy of Sciences.

Tversky, A., & Kahneman, D. (1982). Judgement under uncertainty: Heuristics and biases. In D. Kahneman, P. Slovic, & A. Tversky (Eds.), *Judgement under uncertainty.* New York: Cambridge University Press.

Waddell, C. (1989). Reasonableness versus rationality in the construction and justification of science policy decisions: The case of the Cambridge Experimentation Review Board. *Science, Technology, and Human Values, 14,* 7–25.

Wright, S. (1994). *Molecular politics: Developing American and British regulatory policy for genetic engineering, 1972–1982.* Chicago: University of Chicago Press.

Wynne, B. (1996a). May the sheep safely graze? A reflexive view of the expert-lay divide. In S. Lash, B. Szerszynski, & B. Wynne (Eds.), *Risk, environment, and modernity: Towards a new ecology* (pp. 44–83). London: Sage.

Wynne, B. (1996b). Misunderstood misunderstandings: Social identities and public uptake of science. In A. Irwin & B. Wynne (Eds.), *Misunderstanding science? The public reconstruction of science and technology* (pp. 19–49). Cambridge: Cambridge University Press.

Contributors

STEVEN EPSTEIN is Associate Professor of Sociology at the University of California, San Diego. He is also affiliated with the Science Studies Program, an interdisciplinary graduate program at UCSD. His book, *Impure Science: AIDS, Activism, and the Politics of Knowledge* (University of California Press, 1996), is the recipient of the C. Wright Mills prize of the Society for the Study of Social Problems and the Robert K. Merton prize of the Science, Knowledge, and Technology Section of the American Sociological Association.

SANDRA HARDING is a Professor of Philosophy in the Graduate School of Education and Information Studies at the University of California at Los Angeles, where she also directs the UCLA Center for the Study of Women. She is the author or editor of ten books on issues in the philosophy of science, epistemology, and feminist, multicultural, and postcolonial epistemology and science theory. Her most recent book is *Is Science Multicultural? Postcolonialisms, Feminisms, and Epistemologies,* (Bloomington, IN: Indiana University Press, 1998.)

NEVA HASSANEIN is an Assistant Professor of Environmental Studies at the University of Montana. She received her doctorate in environmental studies from the University of Wisconsin—Madison in 1997 and is the author of *Changing the Way America Farms: Knowledge and Community in the Sustainable Agriculture Movement* (1999), University of Nebraska Press. Neva formerly worked on pesticide policy reform for the Northwest Coalition for Alternatives to Pesticides, a nonprofit organization in Eugene, Oregon.

LOUISE KAPLAN, Ph.D., ARNP, is an assistant professor in the School of Nursing at Pacific Lutheran University in Tacoma, Washington. She has served as the Research Coordinator for the Hanford Information Network, which provides information to the public and health care providers on radiation health effects related to the radioactive releases from the Hanford Site in Washington State. She currently serves on the Hanford Health Effects

Subcommittee, which advises the Agency for Toxic Substances and Disease Registry and the Center for Disease Control and Prevention on public health activities and studies related to the Hanford Site. She is a family nurse practitioner and has a special interest in health policy.

DANIEL LEE KLEINMAN joined the Department of Rural Sociology at the University of Wisonsin—Madison as an associate professor in the autumn of 2000. From 1994 until 2000, Kleinman was a member of the School of History, Technology, and Society, Georgia Institute of Technology. His work has appeared in a wide array of periodicals including *BioScience, The Chronicle of Higher Education, Science and Public Policy,* and *Science, Technology, and Human Values,* and he is the author of *Politics on the Endless Frontier: Postwar Research Policy in the United States* (1995, Duke University Press). Much of his current research effort is focused on an ethnographic exploration of how the "world of commerce" shapes university biology.

DANIEL SAREWITZ is a senior research scholar at Columbia University, where he is developing and coordinating the Science, Policy, and Outcomes Project, and working on a book about linking science to social outcomes. He is the author of *Frontiers of Illusion: Science, Technology, and the Politics of Progress,* (Temple University Press, 1996), as well as many other articles, speeches, and reports about the relationship between science and social progress. Prior to taking up his current position, he was the first director of the Geological Society of America's Institute for Environmental Education, where he implemented a range of new environmental, policy, and outreach activities for this 15,000 member professional society. From 1989–1993 he worked on Capitol Hill, first as a Congressional Science Fellow, and then as science consultant to the House of Representatives Committee on Science, Space, and Technology. His responsibilities included federal research policy, international scientific cooperation, and science education. Before moving into the policy arena he was a research associate and lecturer in the Department of Geological Sciences at Cornell University, with research and publications focusing on processes of mountain building and basin formation along active plate boundaries, and field areas in the Philippines, Argentina, and Tadjikistan. He received his Ph.D. in geological sciences from Cornell University in 1986.

STEPHEN SCHNEIDER, a MacArthur Fellow, is a professor in the Department of Biological Sciences at Stanford University. He is the author of *The Genesis Strategy: Climate and Global Survival* (with L. E. Mesirow, New York: Plenum Press, 1976), *Global Warming: Are We Entering the Greenhouse Century* (San Francisco, CA: Sierra Club Books, 1989), and *Laboratory Earth: The Planetary Gamble We Can't Afford to Lose* (New York: Basic Books, 1997).

RICHARD E. SCLOVE, author of the award-winning book *Democracy and Technology* (Guilford Press), is the founder and research director of the Loka Institute, a nonprofit organization dedicated to making science and technology responsive to democratically decided social and environmental concerns. Dr. Sclove initiated the first pilot consensus conference organized in the United States, as well as the Loka Institute's initiative to establish a worldwide Community Research Network. He is also the founder of FASTnet (the Federation of Activists on Science and Technology Network), and he has published widely in both scholarly and popular venues, including the *Washington Post,* the *Christian Science Monitor,* the *Chronicle of Higher Education, Technology Review,* and *Science* magazine. He may be contacted via The Loka Institute, P.O. Box 355, Amherst, MA 01004, USA; Tel. (413) 559-5860; Fax (413) 559-5811; e-mail <Loka@loka.org>; World Wide Web <www.loka.org>.

Index

adaptive management, 98
Agency for Toxic Substances and Disease Registry, 80
AIDS, 93, 94
AIDS Clinical Trials Group (ACTG). *See* National Institutes of Health (NIH)
AIDS Coalition to Unleash Power (ACT UP), 19, 24. *See also* AIDS treatment activist movement
AIDS treatment activist movement, 147–149
    buyers clubs, 19
    dangers to good conduct of science and, 25, 26
    expert-lay divide, replication of, 24, 25
    goals of, 18, 19
    organization of, 17, 18
    Project Inform, 19
    tactics of, 20, 21
    unintended effects of, 26
AIDS Treatment News, 16, 19
American Association for the Advancement of Science (AAAS), 4, 5
antinuclear activism
    citizen participation and, 77–80
    Coalition for Safe Energy (CASE), 71
    Freedom of Information Act and, 73, 74
    Hanford Education Action League, 74–76

Hanford Oversight Committee, 73
    Initiative 325, 71
    Initiative 383, 72
    Initiative 394, 72
    Yakama Indian Nation and, 72
Atomic Energy Commission (AEC), 70, 76. *See also* Cold War

Barnes, I., 145
Battelle Pacific Northwest Laboratories, 78, 79
Benson, A., 74
biomedical research, inequality and, 93, 94
Branscomb, L., 5
breast cancer activism. *See* National Breast Cancer Coalition
Breyer, S., 161n
Bricmont, J., 2, 6. *See also* Sokal, A.
Bush, V., 3, 90

Cambridge Massachusetts Laboratory Experimentation Review Board (CERB), 144, 145
Carter, J., 73
Centers for Disease Control. *See* Hanford Environmental Dose Reconstruction Project; Hanford Health Effects Subcommittee; Hanford Thyroid Disease Study
Christianity, science and, 126, 132

171